Introduction to Engineering Technology and Engineering

Val D. Hawks
Brigham Young University

A. Brent Strong
Brigham Young University

Upper Saddle River, New Jersey
Columbus, Ohio

Library of Congress Cataloging-in-Publication Data

Hawks, Val D.
Introduction to engineering technology and engineering / by Val D. Hawks and A. Brent Strong.–1st ed.
p. cm.
ISBN 0-13-852402-5
1. Engineering students--Fiction. I. Strong, A. Brent. II. Title.

PS3558.A82319 I58 2001
813.'6–dc21 00-025677

Vice President and Publisher: Dave Garza
Editor in Chief: Stephen Helba
Acquisitions Editor: Debbie Yarnell
Production Editor: Louise N. Sette
Production Supervision: Lithokraft II
Design Coordinator: Robin G. Chukes
Cover Designer: Rod Harris
Cover art: ©Kathy Hanley
Production Manager: Brian Fox
Marketing Manager: Chris Bracken

This book was set in ZapfCalligraphy by Lithokraft II and was printed and bound by R. R. Donnelley & Sons Company. The cover was printed by Phoenix Color Corp.

10 9 8 7 6 5 4 3 2 1
ISBN: 0-13-852402-5

To Julie and Margaret. Our companions and best friends.

PREFACE

Most introductory textbooks for engineering and engineering technology discuss the key aspects of these professions using traditional methods such as charts, graphs, and text. This book is different! It presents these concepts in the format of a story.

Sometimes the most powerful teaching tool is an example. We have found that when students can observe or identify with an example, their learning and retention increase tremendously. Sadly, the opportunity to develop such relationships in learning is rare and difficult to set up in normal teaching environments. We therefore propose a method to accomplish just this type of learning. The method is a novel about a young man who is contemplating entering engineering or technology. We capture the story of this young man's life as he enters college and watch how he interacts with friends, faculty, parents, and fellow workers at his part-time job as a trainee.

We believe the students who read the book will strongly identify with the characters in the story. This story provides the students with a personalized viewpoint that allows them to see what life as an engineer or technologist will be like. They will be able to see how the skills taught in engineering and technology programs and in other parts of their education work together in a real-life setting. This unique approach can have a powerful impact on the perceptions of the students who enter engineering or technology as a career and may help develop characteristics that will guide them throughout their careers. The establishment of this type of paradigm (the view of a practicing technologist or engineer) may be the most important single factor contributing to the success of future engineers/technologists, because they will follow the examples they have been shown in the story. Furthermore, memory retention of the basic concepts is much better because the material is learned in context, thus reinforcing the ideas with other data and relevant discussion.

Most of the basic concepts that are important for an engineer/technologist are presented in the story. These are treated briefly in most cases, but the core concepts are well communicated with sufficient depth so that students can understand the basics and, more importantly, see how to practice the concepts in their lives. A partial list of the concepts introduced in the story include the following:

- What is an engineer and what is a technologist?
- Problem solving using the scientific method
- History of engineering and technology
- Spectrum of science, engineering, engineering technology, and technology
- Math as the language of science

- Different types of engineers and technologists
- Working in a team
- Goal setting
- Study skills
- Importance of general education/lifelong learning
- Effective writing
- Business economics
- Keeping of a lab (work) notebook
- Simultaneous engineering
- Project management
- Charting methods and elementary statistics
- Importance of caring for the environment

When a topic is so large that supplemental materials are needed, these materials are provided in Part Two and the Instructor's Supplement. Part Two is a simulation of a student's notebook, thus giving a further example of how supplementary material should be preserved and reviewed. These are notes that the students can refer to during and after lectures and can also be used by the instructor as outlines for class lectures and discussions. The Instructor's Supplement provides a sample course outline with activities and assignments that have been used in the authors' own classes.

Three main audiences are envisioned for this book. The first is, of course, the freshman in an introductory engineering or technology course. The second is the student, either in high school or in college, or parents of a student who has not yet chosen a major and is anxious to understand what the life of an engineer/technologist is like. The third audience is practicing engineers who may wish to read a book that gives a broad perspective of their occupation.

This book is certainly not the first to attempt to convey technical issues in a novel genre. However, we believe that this book conveys some important concepts and offers a unique insight into the fields of engineering and technology.

Acknowledgments

Any significant effort is not accomplished in isolation. This book is no different and many have contributed in a variety of ways. First and foremost our thanks go to our wives. Julie and Margaret have been wonderful. We appreciate their love, support, and encouragement, but even more we are grateful for the example of commitment, dedication, and excellence in their own chosen careers and life's work that not only inspires us but gives us a target to reach for in our own lives. Our children deserve thanks as well because some of the characters in the book have taken on both their names and attributes. Our wives and children also provided excellent comments and critique of early drafts.

Second, we are grateful to the students of the Manufacturing Engineering and Engineering Technology Programs at Brigham Young University. Teaching the Introduction to Engineering Technology and Engineering class has not only provided the incentive for this book but much of the material. We have a great deal of admiration and respect for these students, who we feel are the best in the world. Their commitment to excellence, their ideas, and their examples are embodied in the person we call Drew Barnes.

Next, a big thanks to the great people at Prentice Hall, Lithokraft II, and others involved. They have been absolutely delightful to work with in accomplishing the details of the book, of course without which there is no book. Their professionalism has been admirable but even more impressive has been their kind way and quick response in every step and on every issue with thoughtfulness and integrity. There is always danger in mentioning specific names, as no doubt someone will be left out. Still, we are grateful to Steve, Debbie, Megan, Robin, Louise, Daniel, Chris, Linda and the many who have worked with them. A special note of thanks to Steve and the others at Prentice Hall for taking a chance on a different format and approach to a textbook.

We would also like to thank the reviewers of this book: Leon L. Copeland, Ph.D., University of Maryland Eastern Shore; and Susan Ramlo, University of Akron.

Finally, we thank each other. We have been colleagues for many years and seen a lot of changes in our own lives, our families, and our careers. The mutual support has been wonderful. More importantly, we have been good friends. The opportunity to discuss, debate, and learn from each other has been a blessing to our lives for the past fourteen years. We look forward to many more.

Contents

Part One: The Story 1

- Chapter 1 3
- Chapter 2 5
- Chapter 3 10
- Chapter 4 21
- Chapter 5 38
- Chapter 6 44
- Chapter 7 49
- Chapter 8 60
- Chapter 9 69
- Chapter 10 78
- Chapter 11 87
- Chapter 12 97
- Chapter 13 102
- Chapter 14 112
- Chapter 15 122
- Chapter 16 134
- Chapter 17 145
- Chapter 18 151

Part Two: Drew Barnes—Reflections, Notes, Handouts, and Assignments from Classes 155

- History of Engineering and Technology 157
- Study Skills and Succeeding at College 168
- Goal Setting and Personal Development 177
- Problem Solving and the Design Process 184
- Practice of Engineering and Technology 191
- Teams and Teamwork 198
- Mathematics and Statistics 205
- Engineering Economics 214

Communication Skills **222**
Creativity and Innovation **231**
Ethics and Professionalism **238**
Company Organizations and Corporate Economics **245**
Social and Environmental Issues **251**
Lifelong Learning **256**
Safety **261**

Index **264**

Part One

The Story

Chapter 1

"Drew Barnes is stumped!" Shaun Stevens exclaimed, jumping around in his seat on the passenger side of the car. His dad gives him a stern look as he drives. Pleased at my dilemma, Shaun was right. This one has me stumped. I am usually one who comes up with the answers to these kind of puzzles fairly quickly.

Shaun's father works for an engineering company in sales. He graduated from State University about twenty-two years ago, the same school we are headed for right now. He produced his puzzle a half-hour ago, and I still can't figure it out. Shaun couldn't get it either and had his dad show him. His pleasure comes from knowing that I usually get these pretty quickly and now I can't figure this one out. Even though he didn't get it, the fact that I am stumped is enjoyable to him.

Maybe I'm not concentrating enough. Ignoring Shaun, I look at the puzzle again. There are three figures arranged in single row and my task is simply to determine what the last one in the sequence should be.

?

Staring at the scrap of paper Shaun's dad gave me with the puzzle on it, I still can't see a pattern.

My thoughts are interrupted by Mr. Stevens' voice. "So are you planning on majoring in an engineering discipline, Drew?"

"Well I was," I reply, "but now I'm wondering if I am smart enough. This looks like it shouldn't be that hard."

Mr. Stevens laughs. "Don't let a little puzzle discourage you. It's one of those things that once you see the answer, you'll hit yourself. It is very simple, but getting to the solution of what seems like a simple problem is sometimes very frustrating. Your studies will often be the same way. But you'll do fine."

I take my attention from the puzzle and lean forward from the back seat. "You studied engineering at State U, didn't you Mr. Stevens?"

He nods. "I did. I originally started in a two year drafting program but ended getting the four year mechanical engineering technology degree. I really enjoyed it. It has been a great career for me."

Shaun speaks up. "Dad works in technical sales at Baldwin Pumps. You manage a group there now, don't you Dad?"

Shaun's dad nods. "I went to work for Baldwin twenty years ago after I graduated. I worked in various areas in the engineering department. Six years ago I took on this technical sales area. We give technical support to the sales and marketing of our product, both before and after a sale to a customer. The group I manage is about fifteen people."

Just then Shaun points out a golf course where he and some friends went golfing last summer. I return to my puzzle but not with full concentration. I find myself thinking about how stumped I feel right now on my way to college. I've leaned toward studying something in engineering, but there are many options at State U. Two year and four year engineering technology degrees, traditional engineering programs, and even a community college in the same town that offers vocational degrees and certifications. Shaun's right, I'm stumped all right. And not just with the puzzle.

Mr. Stevens speaks up again. I look up and see him glance in the rearview mirror at me as he talks. "Drew, how's your dad doing?"

"Pretty good," I tell him, although really I'm not sure. Dad's health has not been good. That's the reason I'm riding with Shaun and his dad instead of Mom and Dad taking me to school. I can tell Mr. Stevens would like to know more, but he doesn't ask and I don't feel like saying more. Dad's health is just one more problem I worry about, but Mr. Stevens doesn't push the issue.

Shaun turns around and asks, "Figured it out yet, Drew?"

I pause for a minute wondering if he means my confusion about a major. Then I see him looking at the puzzle I'm still holding.

I shake my head. "Nope. I'm as confused about this as I am about my major."

Mr. Stevens chuckles again. "Don't worry about either of them. To solve problems you need to be able to concentrate. You have a lot on your mind right now. Heading off to college away from home is a big jump. It's a scary and exciting thing. Hang on to that paper. You'll figure it out."

"Want me to show you, Drew?" Shaun asks. He knows the answer and it's driving him crazy not telling me what it is. I consider his offer but decline. I like to figure these things out myself.

Mr. Stevens speaks up again. "You know where Shaun is going to be living on campus. You can stop by and ask if you don't get it figured out. Just remember," now he is looking over at Shaun as he talks, "to learn and solve problems you have to have time to concentrate without other distractions." He emphasizes the last part of the sentence.

Shaun looks at him and says, "I know Dad, I will." It's pretty clear this discussion has gone on before.

By this time we are at the exit from the interstate to State University. I'm too anxious to continue with the puzzle. Folding it up and putting it in my pocket, I look ahead and see the large clock and bell tower that is the hallmark symbol of State U's campus. Before I know it, I'm bidding goodbye to Shaun and Mr. Stevens as they drive away. I pick up a suitcase and my duffle bag, leaving the boxes sitting on the sidewalk for my next trip into the dorm.

The anxiety I was feeling coming up here from home has been replaced by excitement. Here I am. State University's newest freshman.

Chapter 2

I sit looking out the window of my new dorm room. "Not a bad view," I think to myself. "A basketball court, the Head Resident's parking spot, a couple of trees, another dorm, and a fast food joint across the street." I consider the advantages of the view; knowing when the Head Resident is gone, getting into a good pickup game of b-ball anytime I want, and being able to pick up some pizza without walking all over town.

My first year here at State University, and I'm not sure what to expect. Not to worry though, I'm not really too far from home. It is only a three hour drive. I figure it's close enough to get home when I am homesick (but I don't want Dad to know I might get homesick), yet far enough away to keep Mom from dropping in without a good reason.

Suddenly the door bursts open, interrupting my moments of reflection, and a tall, thin, dark-haired guy yells, "Hey, he's already here." He looks at me now, " Hi, I'm Jim. Jim Stuben to be exact. My friends call me Stu. Since we are going to be roommates, you can call me Stu ."

This guy reminds me of a salesman in an electronics store. Immediately they become your friend. He gives me that kind of an impression, but I've only known him for 30 seconds. I better not make too quick of a judgement.

"Hi," I reply. "I'm Andrew Barnes. I go by Drew most of the time, except when Mom is mad, then I go by my whole name." At that point Stu's mother and father come around the corner. I jump off my bed and politely stand.

"I'm Jim Stuben," Stu's father introduces himself, "this is my wife Rachel."

Stu interrupts, "That's another reason I go by Stu" he explains, "so people can tell us apart. Dad's a jr. and I'm James Archibald Stuben the third."

Stu's mother smiles, "Stu and Drew. I hope people who call don't get you two mixed up."

Stu laughs, "Hey as long as they're female, I'll take his calls anytime."

Stu's dad smiles, then takes over the conversation again. "We're from Lakeside, just a few minutes from here. Where are you from?"

"Springdale," I respond. "It is about three hours away."

"Yes, I know the place," says Mr. Stuben, "I knew a man from there a few years ago. A John Smithson, he runs an insurance agency. Do you know him?"

"I know a couple of Smithson's from high school. They might be related," I respond. I actually knew of them, but didn't know them well.

About this time I notice a pretty girl in the doorway. Stu sees her at the same time. "This is my sister, Kathryn, but we call he Katie," Stu says.

She nods politely. "Nice to meet you," she says quietly. I like her immediately. She seems more shy and quiet than her brother.

Stu's family and I exchange a few pleasantries, then I decide to leave them alone and use going to buy books as a reason to excuse myself.

The campus is nice this time of year with a lot of late summer flowers in bloom and well manicured lawns. As I start up the hill to the main campus, I notice a bulletin board with all kinds of flyers on it. It looks as if there is something on the board from every group on campus and then some. Three particularly catch my eye.

First, a flyer for a campus ski club, one for a sky diving club, and a notice about an open house in the Engineering and Technology college. There are slips with the club announcements, and I take one of each. I make a mental note that the open house is next Thursday evening, and decide I want to be there. It might be a good place to find out some information on what field of study to declare as a major. I enjoyed high school science, and I'm OK at math. Most of all, I like knowing how things work. With my kind of interests and talents, our high school counselor suggested I might enjoy a career in engineering and technology.

There are a few people on the walks. Most are students, but some families are checking out campus. It's a very nice day, not too hot. Late August here can sometimes still be fairly warm. I decide not to head to the bookstore just yet, instead, I'll find the buildings where I will have my classes and see where the rooms are. I like to know where I am going before I get there if at all possible. Besides it will save me time next week when classes start.

First I head to the engineering building. The only class I have here is the Introduction to Engineering and Technology class. It's held twice a week on

Tuesday and Thursday mornings. I enter the building and gradually find the room on the second floor. I try the door to see if the room is open, and it is. I open the door and peak inside. It's dark except for one light in the back.

Suddenly I hear a voice behind me, "Can I help you?" The unexpected question startles me. I wasn't expecting anybody and sneaking a look at the room put me off guard. I turn around to see a middleaged gentleman. "I'm sorry," he says, "I didn't mean to startle you. I just wondered if I could help you with something?"

"Oh, that's OK" I reply, "I was just checking the rooms for my classes. I hope that's OK."

He's still smiling from scaring me. "That's fine. What class is it you will be attending?"

"ET 101," I tell him. "The Introduction to Engineering and Technology course. Do you know about it?" I ask.

He smiles again. "Yes, I do. In fact, I teach it." He holds out his hand. "I'm Professor Stromberg. What is your major, or don't you know yet?"

I shake his hand. "I'm pleased to meet you Professor Stromberg. My name is Andrew Barnes. I'm really looking forward to the class. As for a major, I'm not sure yet which to choose. One that pays real well. I hope I will have a better idea after taking the class."

Professor Stromberg smiled when I mentioned the pay. He nods his head. "I hope so too. They all pay pretty well. That's one of the benefits of choosing engineering and technology. I hope you find more than good pay. Well, Andrew, I'll see you next week. Come ready to learn." He starts to walk away then stops and turns around. "Oh by the way, I start on time, I hope you do. Also, as part of the homework for the class, I like to have all the students involved in a project that helps them learn about the principles and concepts of engineering and technology. You might be giving some thought to what you want to do. If you can't find one on your own, I will have some to suggest. Well, again, I'll see you next week."

Just as he turns to leave again, I stop him. "Dr. Stromberg." I hesitate a little because I'm not sure how to ask my question. "Ummm, do you know any companies that hire students to work around here? I mean in this field, or do they want experienced people? I'm just wondering. I need to get a part time job, and my dad thought it would be good to try to get one that would give me experience related to my studies."

He takes a step or two toward me and responds. "Good idea, but I'm not aware of any jobs right now. Some of the local firms hire interns, but that's usually during the summer months. Most of them will be finishing up at this time." He thinks for a minute then continues. "You might try the companies just south of town in the industrial park. There is an electronic circuit board fabricator called Advanced Electrical Technologies, a plastic injection molding company called Space Age Plastics, and a company that makes actuators and solenoids called Advanced Component Technologies Corporation. I don't know if they are hiring any part time help or not. Try them. Good luck, Andrew."

As he walks away, three things are pretty obvious to me. First he doesn't like students to be late to class. Second, he would rather we come up with our own project than assign us one, and third, the next time I'm going to peek into a classroom, I had better look around first. I've still got the jitters from him scaring me. I'm not sure what to do about a job, but at least he had some leads for me. I flipped burgers at home after school. I guess I could still do that.

I check out a couple of other classrooms then decide to head over to the bookstore. I didn't get to the physics classroom, but I did walk by the building. I figure it wouldn't be too hard to find the room, so I'll do it later.

I was able to get all my books except for the ET 101 class. I couldn't find one listed for it. I wish I would have asked Professor Stromberg when I saw him, but I'll just not worry about it until next week. The load of books is heavy and I forgot to bring my backpack with me, so I decide to head back to the dorm to unload them.

When I walk into the dorm lobby I see Katie, Stu's sister, sitting on the lobby sofa. I say hello to her and she smiles. "Are you getting ready to head back home?" I ask.

"In a few minutes. Mom and Dad went with Stu to the main office to take care of some business. When they get back we'll head home." She looks at the bags I'm carrying. "Looks like you got your books."

"Most of them. Couldn't find a book for one class but I did get all the others. I hope they are as interesting as they are expensive. If they are I should learn a lot."

She laughs then moves over and invites me to sit down until her mom and dad get back. I oblige. "Mom told Stu that maybe you would like to come out for dinner sometime since you aren't very near your own home. Mom's a very good cook."

"That would be great. Who wouldn't like a break from dorm food once in awhile? Although, I hear the food here isn't that bad." As I unconsciously take the piece of paper with the puzzle on it from my pocket, I ask, "Your brother seems very outgoing. Is he like that all the time?"

She laughs again. I like her laugh. "With Stu, what you see is what you get. He is outgoing, people like him, and he is good at school and sports." She's looking at the paper with the puzzle I'm holding. "What's that?" She asks.

I realize I've been looking at it as I talk. "Oh, it's a puzzle a friend gave me this morning on our ride here to school."

"I like these kind of things. Can I see it?," she asks as she reaches for the paper.

I give it to her. She holds it so we can both see it and says nothing for a couple of minutes. Just concentrates on the figures on the sheet.

Suddenly she says. "Oh, I see it. Wow, that's really neat." She takes the pencil I've also been holding and sketches the last figure. I watch her draw in a 1, then two 0's connected to each other, then another 1, except backwards, beside the 0's. I look at it wondering how in the world she came up with it so fast. I now see the other figures are the numbers 1, 4, and 7, each mirrored. The next number in the sequence is the last number plus 3, drawn mirrored to itself. Mr. Stevens was right. When you see the pattern it is very obvious, and I almost hit myself wondering how I missed it.

Katie smiles and hands it back to me. "I like that one. Did I do OK? How long did it take you to solve it?"

I smile and answer shyly. "Longer than it took you. You did great. Maybe you ought to come here and study technology or engineering."

She laughs and says. "I plan to next year." Just as she's about to say something else I hear Stu's voice.

"Flirting with my baby sister, huh Drew? Well, good taste my friend, good taste."

Katie blushes a little but takes it right in stride, turns to me and says, "See? That's Stu. What you see is what you get."

Stu tells Katie that her mom and dad are waiting in the car. As she gets up to go she says, "Drew, I hope Stu doesn't drive you crazy. I've been watching out for him for 18 years, now I hope you'll take good care of him." She smiles as Stu wraps his arm around her and gives her a one armed hug.

Stu puts on a sober face and says, "I'll be OK, sis. If I need you, you're only a phone call away."

Katie turns to me again. "I hope you will be able to make it out to dinner sometime. Have a good semester." I thank her for the invitation and assure her I'll be out. She waves again as she goes out the door.

I grab the bags of books and walk to the room with Stu.

"Well my friend," Stu says as he takes one of the bags, leaving me with the other. "I guess we are real college boys now, so we should act like it." He looks stoic for a moment then breaks into a big grin. "Hey, I know! Let's order some pizza."

This is going to be quite a year.

Chapter 3

I feel a little nervous as I sit waiting for the first lecture in the Introduction to Engineering and Engineering Technology class. The attendance is smaller than I expected. Obviously the stories I have heard about hundreds of students in every class won't hold true in this one. Like a typical lecture hall in its shape, narrowing toward the front and inclined upward toward the back, it differs in that there is carpet and cushioned chairs. "I could get some rest in these chairs," I think to myself as I select a seat in the middle of the room. I don't want to be too close or too far away. I figure I'll take stock of the professor and the class before I make a choice where to sit in the future. The room is filling up quickly now and when class starts it's about 3/4 full. My guess is there are 60 to 70 people in the class.

I recognize four guys from my dorm. One of them, a guy named Russell, also recognizes me and sits down by me. He is about 4 inches taller than me, and skinnier too. He's the kind of guy that, if he turns sideways and sticks out his tongue, looks like a zipper.

"I didn't know you were interested in engineering," he says as he makes himself comfortable. Without waiting for a response he asks, "What kind of engineering do you plan on studying?"

"I'm not sure. That's where I hope this class will help me out. I'm hoping to learn about different kinds of engineering to help me make a decision. By the way, I'm Drew," I tell him, not sure if he knows me.

"Yes, I know," he replies. "I'm Russell. You live in my dorm. Maybe we can study together."

"That would be great," I tell him, hoping that he is smart enough to help me out, but not so smart that I'll feel dumb.

The room quiets a little when the professor walks in but there is still some talking going on. Russell and I don't talk. I just watch the professor. He stands at the front of the room looking out over the class, not saying anything. It starts to become obvious that he is waiting for the class to quiet down. This may take a while I think, but about that time I hear a few "sshhhh" sounds and soon every one is looking at the professor. I get the feeling I'm not in high school anymore.

As if reading my mind, the professor starts to talk. "You are not in high school anymore," he starts out. "You are here because you want to be and you are here to learn. I don't expect to have to quiet you down when I come in. You are welcome to visit quietly until I begin talking, then I expect you will come to order." He doesn't speak loudly or harshly, but I get the feeling he means what he says. "Also," he waits for a moment as the door opens and a student sheepishly walks in and sits down in an empty seat near the door. I feel bad for her. "I expect you will be here each day, and be on time." Now I really feel bad for her even though the professor didn't change his voice or look at her. "I am very sorry," I hear the tardy student say. I admire her courage. The professor doesn't say anything but smiles slightly and nods at her.

He continues, "I am Phil Stromberg. I will be your instructor for this class and look forward to getting to know you. You may call me Dr. Stromberg or Professor Stromberg, it doesn't matter. This course is Introduction to Engineering and Engineering Technology, or if you are looking at your class schedule, it is ET 101." Even though I know I'm in the right place, I find myself looking at my class schedule to make sure.

Professor Stromberg then describes the purpose of the class, all of which is why I am here, and proceeds to explain the grading requirements. Most important (at least he says it is), he hopes we will try to forget about the grade and concentrate on learning something. "Easy to say," I think to myself. "I'll bet it is hard to do, especially when it comes to the end of the semester."

There is some guy sitting right up front, writing like mad. He has all kinds of questions. He even asks if the class purposes Professor Stromberg gave earlier are going to be on the test. Professor Stromberg looks at him kind of funny but only says, "It might be." Now the guy writes even faster. I've seen this kind before, I already know he will ace the class or go crazy trying.

After Professor Stromberg asks again if there are any questions, and no one asks, he opens a notebook in front of him. He asks, "What is engineering?" He looks around as if to expect an answer. A few seconds pass and when no one answers, he asks again, "What is engineering?"

Finally a girl in front of me says, "Designing things."

"Good, designing things" Stromberg repeats her answer as he writes it on the board. I'm not sure if I should write it down or not. Finally, I decide I will.

"What else?" He is still at the board looking at the class.

Someone else volunteers, "Problem solving." He writes this one down too.

Another voice calls out, "Creating things." This goes on the board also. Professor Stromberg looks out over the class waiting for more answers.

Another comes, "Doing experiments."

As Stromberg writes that one, a guy behind me says, "My dad is an engineer and he says he writes a lot of reports." "Writing reports" is listed on the board.

Finally Professor Stromberg speaks again, "Good list, but look at it. Is this really what engineering is?" I look at the same list I have written in my notes.

What is engineering?

1) designing things
2) problem solving
3) creating things
4) doing experiments
5) writing reports

It looks OK to me. But I admit I hope the report writing is not true, even though I did well in high school English.

Professor Stromberg continues, "What do you see that is interesting about this list?" We all look at it. I know he is looking for another answer, but I don't know what. I look at my list again. A thought hits me. These things look to me like the things that an engineer does. Maybe that is what he is after. I lean over and whisper to Russell, "These are things an engineer does, not necessarily what engineering is."

"Raise your hand and tell him that," Russell urges.

"No way," I tell him.

"Then I will," Russell responds, as he raises his hand. Professor Stromberg sees him and calls on him. Russell says, "These look like things an engineer does, but maybe that is not the same as what engineering is."

Professor Stromberg looks impressed and says so. "Very good. What's your name?" he asks.

"Russell Stanford sir"

"Very good Russell," he says. Then asks, "Does anyone have a dictionary?"

Yea right, I'm thinking to myself, who would carry a dictionary with them to class? Just then, the "Is this going to be on the test" guy in the front row says "I do," and pulls one out of his backpack. I can't believe it.

"Would you please open to the word engineering and read it to us?" Professor Stromberg walks back to the board. Mr. Dictionary looks it up, then he reads

> *"Engineering is the application of science and mathematics by which the properties of matter and the sources of energy in nature are made useful to people in structures, machines, products, systems, and processes."* *(Webster's New Collegiate Dictionary)*

Dr. Stromberg writes the definition beside the list given earlier. He then underlines the words *application, properties* and *useful.* "This is what engineering is," he says pointing to the new definition. "This" pointing to the other list, "just as Russell said, is some of what an engineer may do." Russell leans over and tells me thanks.

Pulling an overhead transparency from his folder, Stromberg flips on the overhead projector and says, "Let's take a closer look at this definition and what it means. There are three very important aspects of this definition that are the founding principles of good engineering."

On the overhead are the underlined words with some explanation. I copy it down in my notes.

1) Application of science and mathematics. — [having to do with basic natural laws.]
2) Properties of matter and energy. — [principles of behavior in the world around us.]
3) Made useful to people. — [combining the first two for the benefit of society.]

Professor Stromberg is saying something about the three points, but I'm not listening very well. I'm thinking instead. As I look at these three underlined words I realize I can almost make a sentence out of the three themselves. I jot it down.

Application of properties made useful.

I add some of the words from the explanations on each line to make my definition more complete.

> Engineering is the application of basic natural laws and properties of the physical world around us to make useful products for the benefit of society.

Maybe it is because I have never thought about it before, but I am intrigued at the common sense of the definition. I like how it includes the idea of making things that are useful to society. About this time, I realize that while I am thinking, Professor Stromberg is still talking. I hear him finish up a thought. "Then later on in the class we will dabble with what is meant by properties of matter and energy. We will also investigate how, in application, these translate to being useful to society. Let's turn to one more matter for today." He takes the overhead off and erases the board before he continues.

"This is a class in the introduction of engineering and engineering technology. We have programs for both engineering and engineering technology here at State U. I want to discuss the difference between the two because we have students from both programs in the class." He pulls another transparency from the folder but just holds it while he continues to talk. "The difference is much more apparent in academics than in the working world. The basic tendencies and characteristics between engineering and engineering technology are chiefly due to differences of focus. The academic programs here try to capitalize on those differences."

He puts the transparency on the projector. "If we think of science, engineering and technology as a spectrum, we take a broader view of it. We look at it with pure theory on one side to strict techniques on the other." The graph he uses shows what he means. "This is how engineering and engineering technology would fall on the spectrum."

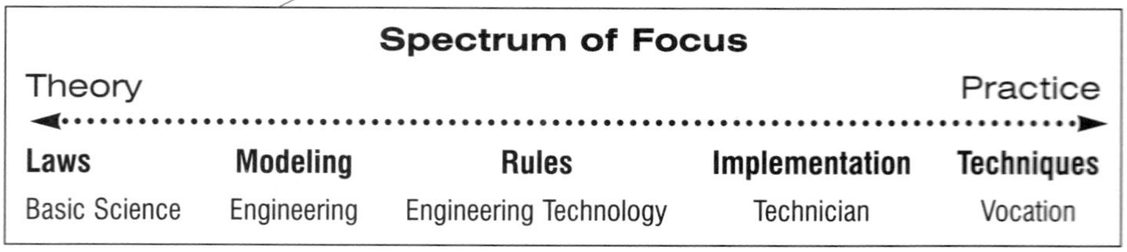

Spectrum of Focus

Theory ◄······► Practice

Laws	Modeling	Rules	Implementation	Techniques
Basic Science	Engineering	Engineering Technology	Technician	Vocation

Stromberg continues his description. "As you approach the purely theoretical side you need more math and basic science to understand and describe basic physical and natural laws. Therefore you will see the engineering programs with about twice as much math and physics as our engineering technology programs, even though both are four year degree programs." Professor Stromberg turns off the projector and as he continues to talk jots a few things on the board.

"The engineer and the engineering technologist are quite similar in some respects, but have different inclinations and expertise to apply to problems and ideas. As I said, the background of the engineer includes more math and physics. Therefore, the engineer has the inclination and ability to develop sets of rules by what we call analytical modeling. Analytical modeling uses science and math." I record what he has written on the board in my notes.

Engineer

—more math and physics
—analytical modeling
—develops rules sets

He keeps writing on the board while he talks and doesn't see Russell raise his hand. When he does turn around, he glances down at his notes, so Russell speaks up. "Professor Stromberg?"

Stromberg looks up, sees Russell's hand up, and asks, "Do you have a question?"

"Yes sir. You mentioned math and the different amounts of math that engineers and engineering technologists will take. I guess I have two questions. First, why is math so important, and second, why calculus?"

"Good question. Every field of study has its own jargon and terms. In that respect, engineering and engineering technology are no different. But even more basic than jargon and terms is the language of engineering and technology. The elements of this language allow communication across disciplines and specialities. The language of engineering and technology includes skills in mathematics, visual representations, and formulas. But let's just talk math for a minute."

Dr. Stromberg must have been planning on this because he goes back to his package of overheads and takes one out while he continues to talk.

"You ask, 'Why math?' Let me share a statement by Alfred North Whitehead, who spoke of the role of math in the history of thought." He places the overhead with a quotation on it on the overhead projector, turns the projector on, then reads the quote while we follow along.

> I will not go so far as to say that to construct a history of thought without profound study of the mathematical ideas of successive epochs is like omitting Hamlet from the play which is named after him. That would be claiming too much. But it is certainly analogous to cutting out the part of Ophelia. This simile is singularly exact. For Ophelia is quite essential to the play, she is very charming, — and a little mad. Let us grant that the pursuit of mathematics is a divine madness of the human spirit, a refuge from the goading urgency of contingent happenings."
>
> Alfred North Whitehead, *Science and the Modern World*

Stromberg continues, "You may not be too familiar with the play *Hamlet*, but I assure you that, as Whitehead states here, an essential element is lost without Ophelia. It is the same with mathematics in engineering and technology. We may still talk about it and even use words to explain it. But precise detail is lost and, without mathematics, is unexplainable. Mathematics and analytical methods are the best ways to describe time/space relationships, forces and stresses, derivations of equations, dimensions, units, equations, number systems, and shapes."

I sense that other students are as intrigued by this discussion as I am. I think it may be because many of us have struggled with math and have wondered why something that seems so theoretical can be helpful in actual practice.

A student behind me speaks up. "Dr. Stromberg, I can sure see why some math is important, but some of the high level math, like calculus and stuff, is really confusing. Why can't we just describe things with basic math skills?"

It appears Dr. Stromberg was ready for this as well. "Several reasons come to mind." He writes them on the board as he talks. "First, you get a description with math that you can't get from experimentation. For example, volume or area under a curve could be measured but not as precisely as with mathematics. With math you can also predict a result not easily obtained by experimentation. You also get an ability to analyze phenomena more deeply and completely than without math. You can even develop a mathematical model that allows generalization of a phenomena. This is extremely valuable and efficient. Math provides the basis for documenting and communicating technical information."

Dr. Stromberg continues, "Every language has a basic structure and organization. Arithmetic, algebra, geometry, trigonometry, and calculus are the forms for mathematics. They have the basic properties, elements, and definitions we need in engineering and technology. For instance, with calculus we can speak in infinitely small units which we call derivatives, and very large units which we call the integral. We use that symbolism to make the laws of the universe as described in chemistry and physics more simple. Calculus gives us a much more powerful language than words with which to communicate these concepts, and it is widely recognized as such. Math is the language of the sciences. This is why some schools allow math to be counted as general education language credit."

Dr. Stromberg stops talking for a moment as if to let what he has just said soak in. I knew that math counts as language credit but I didn't know why. It's like a lightbulb goes off in my head. While I am pondering, another student asks a question.

"Dr. Stromberg, why do we also need to take math beyond calculus?"

Dr. Stromberg smiles. "Another very good question. Engineering technology and engineering students take calculus. Even many non-engineering students, like those in business and statistics, take calculus to learn the basic concepts. Most engineering students must also take more, which we call engineering math.

"Let me explain it this way. How many of you have studied a foreign language?" It looks like over half the class raises their hands.

"Good," Dr. Stromberg continues, "Now how many of you would say you speak the language fluently or at least can converse reasonably well in the language."

About half the hands go down leaving approximately a quarter of the hands still up.

Stromberg asks again. "Now how many of you feel confident talking to someone about politics with all its jargon and special terms in the language?"

I look around and see only one hand up now. Dr. Stromberg asks the student with the hand still up, "How is it that you learned to speak that well?"

The student replies, "My dad works for an automotive company in Mexico and I grew up listening to him speak with others about political things in our home. It was part of my learning to speak Spanish."

"Good," Dr. Stromberg explains. "The basics of mathematics are the elements of conversational language for engineering and technology. Anyone studying the field needs this background. More classes in engineering math provide additional proficiency for more powerful analysis and modeling, similar to learning the jargon of politics or religion in a foreign language."

Dr. Stromberg looks at Russell. "That was a rather long answer, but was it helpful?"

"Yes sir," Russell declares. "Now I feel like I'd better take my calculus more seriously. I don't want to be a tourist in the land of engineering and technology."

Everyone chuckles and Dr. Stromberg says, "Good, I'm glad you got the point." Then he continues. "An inclination and interest in math provides a young engineer with a great ability to communicate in the technical realm of the field. However, this certainly is not sufficient. As with any society, an understanding of the culture and environment is important. But of course, communication is very difficult without the language. I hope all of you will take your math seriously and learn it well. The same is true of the other technical classes you take whether they are civil, mechanical, manufacturing, or electronics classes. You are learning skills and gaining education in methods, but you are also learning to communicate in your chosen area."

Dr. Stromberg turns the overhead projector off and goes back to the board. "Now let's continue our discussion on engineering and engineering and technology."

He starts writing again while he talks. "The engineering technologist has more experience in the actual application of the process, operations, and methods. Often, he or she has an inclination and ability to use existing rule sets in innovative ways to solve problems through technical intuition and experimentation. Often the engineering technologist has a wider view of the business enterprise, while the engineer has a deeper view of the theory. Though engineers and engineering technologists may attack problems different ways, they are often called by similar titles, even the same titles in some cases. They often work in the same organizations at the same kinds of tasks. In common practice, the work of both engineers and engineering technologists are called the 'field of engineering.' Hence, job titles for both are often just 'engineer'." I finish off my notes with what he has written on the board.

Engineer	Engineering Technologists (implementors)
— more math and physics	— actual experience & problem solving
— analytical modeling	— innovative application of rules
— develops rules sets	— experimental and intuitive

The guy next to me leans over and says, "That's why I'm in engineering technology." He points to the list on my notes. "I like to jump in and solve problems and use current technology in innovative ways. I do OK with math, but I don't want to do much analytical modeling. I want to be in the middle of the action, where things get done."

I begin to hope that Professor Stromberg is getting tired because I want to take a rest. A hand is up though, and Professor Stromberg nods at the student to go ahead and talk.

The student asks, "I feel like I am pretty good at science and math, but I don't really know if I should study science, engineering, or engineering technology. Which is the best to study? Isn't it more prestigious to be a scientist than a technician?" I'm glad she asked the question, because I'm wondering the same thing.

Stromberg walks away from the board toward the class to answer. "Those are very good questions. These descriptions of the scientist, engineer, engineering technologist, and technician deal more with academic disciplines and functions than they do with people. In reality, many people who work in the science and engineering fields are a combination of these functions more than they are strictly one or the other. Unfortunately, we too often try to rate the comparative value of scientists, engineers, engineering technologists, and technicians. Too often, we argue about which is best, but no conclusions can or should be made. All these functions, from the technician to the scientist, are valuable and fill important roles. The differences are those of focus and the approach to problems."

Now he has gone back to the overhead projector and put the spectrum transparency back on it. He takes a marker from his pocket and writes *Theoretical Physicist* above *Science* on the left hand side, and *Mechanic* above *Vocation* on the right hand side. Then he continues to talk.

"These two functions may be placed at the ends of the spectrum. They require a different set of talents or a different level of education, but both are critical and serve society very well. To argue about the value of one over the other offers no benefit to either, and detracts from the very important responsibility to work cooperatively." He has now written the word *value* underneath the graph on the overhead and put a big question mark under it.

He looks at the class and asks a question. "Would you rather get help from a physicist or a mechanic?" No one answers at first, but Stromberg says nothing. Finally a student in the back speaks up. "It depends on if I am doing physics or fixing my car."

"Right." I can tell Professor Stromberg is really into this. "If you're describing the theory of laws of motion you would want someone who is blessed with talents in that area." He circles the word "Physicist" as he talks, then continues. "But if you want your carburetor fixed, your best bet is a talented mechanic. Both have a real value to society as long as they are hard working and honest."

The student from the back cracks, "Yeah, but just try to find a hard working physicist or an honest mechanic." Stromberg just smiles and everybody laughs. The student continues, "Really though, isn't the physicist worth more than the mechanic? Isn't he paid more?"

"At a given time, his work may be worth more to a certain task or to a certain organization. But the same can be true of the mechanic." Professor Stromberg continues, "But," he goes to the board, writes a "$" symbol on it and puts an X through it. "Pay is not a very good indicator of the real value of someone's work in terms of benefit to society. Sports and entertainment have proven that." Now he gets the laugh. He continues, "Consider it in terms of the idea of being professional. That's what all of you will be when you graduate. You will be part of a profession with educational certification called a degree. The mechanic has a vocation, and probably makes less money than an engineering or science professional, such as an engineer, engineering technologist, or physicist. But all have value to society. If you want to look at the salaries, however, the engineer and the engineering technologist usually make about the same. Physicists with a BS degree usually make less." This also confirms my thinking that engineering and technology is the place for me to make good money.

The clock indicates class is over and Professor Stromberg quickly adds, "For the next class period, please write descriptions of what a scientist, engineer, engineering technologist and a technician all are. Make it one your Aunt Beatrice could understand. Thank you, and have a nice day."

As I gather my books Russell asks, "Hey, should we get together tonight in the dorm study room? We could write them together." I wonder aloud if that would be OK with Professor Stromberg.

"Tell you what," he says, "I'll ask Stromberg if it is OK, and then let you know. In the meantime, let's each write our own. If it is OK, we will get together and compare ideas. What do you think?" This idea I like.

"Deal," I respond, and Russell heads off to talk to Professor Stromberg. I'm on my way to English class. Better get some English under my belt if I'm going to have to write a lot of reports.

Chapter 4

The first few days of classes went as you might expect. I was late to one because I couldn't find the room. It was the only one I didn't check out last week when I took a walk around campus. But I wasn't the only one late, and the professor didn't even seem to notice. Maybe she was used to people not being able to find the classroom.

I got the general idea from all of my classes that how well I do is up to me. Of course, there are no scheduled parent/teacher conferences like we had in high school. And, though some of the classes use attendance as part of the grade, there is no real threat from the professor if I don't go. I can do as I wish.

I wonder how dedicated I really am, or will be, to my studies. I know myself well enough to realize that I have got to find a way to make sure I follow through with my schoolwork. Even though I may tell myself all kinds of great things about how well I will do, when it comes right down to it, talk is cheap.

Dr. Smithson, the physics professor, suggested we check out the student consultation center and sign up for some of their classes that teach techniques for being a successful student. Study habits, time management, and setting priorities were all some she mentioned that will be important to us. Since it is lunch time, I decide to grab a bite to eat at the cafeteria before I head over to the consultation center. Then I'll head back to the dorm to do some studying.

While I'm waiting in line to get lunch, I take out my physics book to check out the assignment we were given. I am interrupted by a voice behind me in line. "You're in Physics 120 aren't you?" He says it as though he has known me for years.

"Yes," I reply, still holding my book open. "I just came from class with Dr. Smithson."

"I have her for physics too." His expression changes little, but he smiles a little as he says it. "Looks like a pretty big assignment for the first day, don't you think?"

I close the book now and slip it back into my backpack. "Yeah, it does. But it's pretty obvious that these classes aren't going to be like high school. Dr. Smithson seems pretty set on getting us started early. I guess that's OK, I just hope I can handle it. What section are you in?"

"Same one you are." He points ahead and I turn, realizing we are at the food counter. I take a tray and move down the line to order a hamburger. He follows me. "I just came from there. I didn't see you, but it's a pretty good size class. Easy to miss somebody."

He's right. There must be close to 300 people in the class. It is easy to get lost. Nobody would know if you are missing unless they are looking for you. "What's your major?" I ask, my hamburger nearly ready.

"Physics. This is the first required class for majors. What's yours?" He ordered a hamburger too. I guess we both just want to stick with standard fare for today. I tell him I am a pre-engineering major. It's easier than explaining that I am not totally certain, but expect to end up somewhere in engineering or engineering technology. He doesn't look like he is with anyone so I invite him to sit with me. He seems grateful and we find a vacant table.

My pre-engineering strategy doesn't work. "What kind of engineering?" he asks. I explain that I'm not sure but that all of them require this physics class so I decided it would be a good one to take. We just chat for awhile about hometowns, majors, dorm life, and weather.

"So where are you headed now?" he asks as we are finishing up. "Oh, I thought I would check out the student counseling center and find out about those classes on study habits that Dr. Smithson talked about. I figure it wouldn't hurt me to review basic learning skills."

"You know, I thought about it but had decided not to take the time. Maybe I'll tag along though, if you don't mind. I have a class at 3:00, but nothing till then. I was just going to start on the homework."

"Glad to have you along. Maybe we can learn more together. Besides, if I make friends with a physics major, I can increase my chances of a good grade in the class."

He laughs, "I said I was majoring in physics! I didn't say I knew that much about it." I can tell he is just trying to be humble. I'll bet he is really pretty good. At least I hope so.

When we get to the consultation center, we find we need to sign up for short classes that are offered. We both sign up for a "How to be successful in college" class taking place that night at 7:00. We agree to meet at the room where the class will be. We are just about to part when he stops short. "I didn't even introduce myself. I'm Troy Stoddard. What's your name?"

"Drew Barnes," I tell him. "Sorry about that. I forgot too. Well, I'll see you tonight. We'll learn how to be overnight college stars." With that we separate, him to the library and me to the dorm.

When I get back to my room, Stu's there sleeping. His CD player is still playing his favorite CD. I drop my pack on my desk and plop on my bed. Stu doesn't stir. As I lay there for a minute, I realize I could fall asleep too, if I'm not careful. I am about to go ahead and doze off, then I realize that if I don't get started on some homework, I might not get any done. I expect that Troy, the guy I ate lunch with, will have some of the physics homework done and might ask me about it. I don't want to be embarrassed, so I decide to get up before I crash. I can't believe this. Only the second day of class, and I'm worried about getting homework done.

I sit at my desk, but decide it might be best to try the study room in the dorm. There might not be anybody there, and I won't wake Stu. As I walk down the hall the thought hits me that I might have been further ahead to go to the library with

Troy, but it's too late now. It's about a 10 minute walk to the library, and I don't want to take the time. Besides, there is only one other person in the study area. It's quiet right now, and there is lots of room to spread out my books. "This will work," I think to myself and choose a table. I open my book and check the first problem. Something about forces. I go to work.

Just before 7:00, I walk in the room where the "Succeeding in College" class is being held. I look around and see Troy wave at me. He is at a table right at the front of the room. I walk over and put my pack down. "Are you one of those guys that always sits in the front?" I tease him. He fakes a shocked look on his face and points at himself, saying, "Me? I got this table because you looked like the diligent, hard working type who sits in front. I just didn't want to be left in the back alone."

"Yeah, right," I sneer back. "If I'm so studious, what am I doing here?"

He chuckles and asks, "Did you start the homework yet?"

I'm glad I did. "Yes, I got through most of the problems. I need to look at number 7 again. I know Dr. Smithson talked about the stuff in class but I haven't figured it out yet. How about you?"

"I got most of it done too. What was seven about?" As he asks, the class instructor walks into the room. "We'll check on it after class," he says quietly and we turn our attention to the teacher.

The man teaching the class doesn't look old enough to be a professor. He can't be much more than 24 or 25. He stops in front of the class and begins to talk.

"My name is Scott Handsen. I am a graduate student in educational psychology, and I will be your teacher for this short course. I know you are all here because you want to be, because this is on your own time and there is no grade. It will be well worth your time, I assure you."

He walks over to the table which is in the front of the room. "I'm here because how people learn is part of my research for my graduate work. If you have any questions when we are done, please feel free to ask me. If you are willing to participate in a survey and study as part of my research, there is a note card you can put your name on. It will be passed out at the end of the class with the evaluation."

With this he walks to the overhead projector and turns it on. I look around the class. There are only about 12 people in the class even though the room could hold 3 or 4 times that many. I expected it to be more crowded. Scott starts talking again.

"There are a number of factors that will contribute to your success at college. Most factors you have control over and can even improve with a little knowledge, practice and discipline. How you organize yourself, your situation, and your time, and how you handle events that occur will in large part determine your success. Contrary to some popular attitudes, other people, other things, and even your past does not dictate your ability to be successful and happy."

ACADEMIC AND PERSONAL SUCCESS

Persistence and hard work
Discipline
Practice
Knowledge

He puts an overhead transparency on the projector.

He continues, "In this class we will discuss a number of habits to make sure you accomplish the things you desire and that you do them well. Development of the habits and characteristics outlined in this class will not only help you be a better students, but will be beneficial to other areas of your life as well."

"Sounds kind of formal, doesn't he?" Troy whispers. I nod. He does talk formally, but doesn't seem removed from the class. I write down the things from the overhead as Scott continues.

"We will talk about a few important topics related to this idea of success. These are as follows:" He stops talking and begins to write them on the overhead with a marker. I copy them on my notes.

ACADEMIC AND PERSONAL SUCCESS

Persistence and hard work — not giving up and being willing to put in the time necessary to be successful. The most important attributes to success.
Discipline — to make the above work by setting yourself up to succeed and follow through.
Practice — trying them out with an attitude of making them work.
Knowledge — of goals, setting priorities, confidence and esteem, learning styles.

Scott underlines the word *Knowledge,* and then continues, "Understanding these points, then saying to yourself, 'I really want to see how this would work.'" He underlines the word *Practice.* Then, as he underlines the word *Discipline,* he continues, "And then sticking with it so it can work for you," he now turns back to the class, "is the key to being successful."

He now goes back to the point of persistence. "Nothing else will matter if you are not persistent. These two," he underlines *Persistence* and *hard work,* "will mean the difference between long term success in your lives and careers and failure or mediocrity."

Instead of another overhead, he shuts of the projector, sits on the table in the front, and changes to a little less formal posture. "Let's start with goals." He pauses then continues talking.

"You know, I think I'm starting to understand that few things have the same impact on accomplishing long and short term objectives as setting and working toward challenging goals. It's possible to live without goals, but life will mean more if one has positive, challenging, and worthwhile goals. Much more will be accomplished in your life by setting, working on, and achieving goals." He picks up a book he had set on the table when he came in, then continues.

"Goal setting should not be a hassle but a calm, periodic, reasoned review of what we value enough to spend our time with." Holding up the book, now he stands up. "An exercise developed by Alan Lakein, an acknowledged expert on time management, uses an interesting process to set valuable goals that refer to the here and now as well as the future. Let's try it."

Scott moves to the board and writes as he talks. "First, I want each of you to write on a sheet of paper five major lifetime goals you have. These are *major* goals. I expect graduation from college is one, a career is probably another, etc. Major things you expect to accomplish in your life. They can be professional or personal. I will give you two minutes to do this. Go ahead."

"Whoa," I'm thinking, "I wasn't prepared for this. Five major goals for my life." I am kinda lost for a few seconds and not sure what to do, when I decide to just go for it. This could be pretty interesting. I don't get five down by the time Scott tells us to stop, but I did get four. The ones I did write down are:

Major Life Goals — Drew Barnes

1. Graduate from college.
2. Have a successful career.
3. Get married and have a good family.
4. Own a cabin in the mountains.
5.

I felt a little embarrassed about the cabin thing but I couldn't think of anything else quickly. Besides, I would like to have one.

Scott continues, "Now that you have five things you want to accomplish in your life, the next thing I want you to do is write down five major goals you want to accomplish in the next year of your life. Again, you have two minutes."

I think I am starting to see the pattern he is setting up here. I think he wants us to see how the shorter term fit with the long term, but I decide to not try to link them right now. I don't have enough time. I do get five goals down this time.

One Year Goals

1. Pass my classes this year.
2. Learn to sky dive.
3. Decide on a major.
4. Meet some great friends.
5. Learn some good habits to make me successful.

The next assignment by Scott is a little different than I thought it would be. Now he asks us to write down five major goals we would pursue if we knew we only had six months to live. This makes me think differently about what I write. I think that is his purpose, and it slows me down, so I only get four again. Some things didn't seem as important when I thought about having only half a year to live. My list looks like this:

Six months to live — what are five goals

1. Learn a lot this semester in school.
2. Visit Alaska.
3. Learn to sky dive.
4. Get to know Mom and Dad and the family better.
5.

As I look at these I mentally confirm that I still want to do well in school. I don't think spending my last six months partying would do anything for me. I still want to learn stuff, but I'd also get in some things I really want to do, but expect to do later, like see Alaska (I've always wanted to do that) and learn to sky dive. I hadn't thought before about getting to know Mom and Dad better. I figure that if I won't have time to have my own wife and children, I'll spend some time with those I do have. It just seems important.

I thought I had figured out where Scott was heading, but now I'm not ready to take any guesses. He is at the board again and writes down the last thing for us to do.

"Now," he is still speaking as he writes the last instruction, "write down, again in two minutes, specific things you have done in the last week that relate to the goals you listed in three previous lists."

This one does throw me. "Specific things I have done that are related to these goals?" I ask myself. I can't see what I have done at all toward these goals. I glance over at Troy. He's fidgeting with his pencil and just looking at the paper. My guess is that he's as blank as I am.

I reflect back on the last week. What have I done this last week? I quit my job at the burger joint. That doesn't seem to have anything to do with the goals I listed. I moved out of the house and into the dorm. That seems to go against my "six months to live goal" of getting to know the family better. It does put me at school where I can take classes and learn. That, and registering for classes fit together with my learning goal so I write them down. I went camping with a couple of friends. That doesn't seem to link to anything. I also talked to Dad about getting a job with an engineering company close by where a friend of his works. That might help me do well in my career. I know I'm grasping at anything, so I write it down. By the time Scott stops us, all I have is:

Done in the last week toward accomplishing the goals

1. Moved to school.
2. Registered for classes.
3. Got lead for a job.

The last one popped into mind at the last moment and I need something to make my list look longer so I put it down. Besides, as I think about it, it could be one the best things I've done this week as far as setting and achieving the goals.

Scott sits on the table in the front of the room again as he continues. "This exercise tends to focus your mind on those things you feel are of the most value to you and you desire the most. We all tend to spend our time and money on what is really most important to us. Now let's talk a little more about goal setting, then we will come back to the things you wrote down on your lists."

Scott now gets off the table and stands in front of the class. "Goal setting and worrying about time can become obsessive, but it is still important to have some kind of method to plan and test what we are doing about things we say are important to us. We need to set the medium and short term goals as well as the long term. It is possible, for example, to set a very aggressive, yet attainable long term goal, yet not reach what we want, because we lack short term goals that we can accomplish on a weekly or daily basis. By making and keeping these short term commitments, we make things happen. By doing this, we won't find ourselves waiting for things to happen. We are busy making them happen."

Scott gets another overhead from the pile. "Many experts on goals believe all goals must be measurable. Perhaps so. At least many ought to be. It is hard to know if you have made progress if the goal cannot be measured and evaluated."

Scott now takes out a marker and writes two sample goals.

"For example, a goal that says 'I will do better on my calculus assignments this week' will not be nearly as effective as 'This week I will spend two hours a day studying calculus.' One reason is that at the end of each day you will know how you are doing. You can see your progress every day." Scott straightens up from writing on the overhead and continues. "The other thing something like this goal does for you," he's pointing to the studying calculus two hours every day, "is help you develop discipline. Discipline will be a great benefit to you throughout your life in professional and personal endeavors."

"It is harder to measure a goal such as, 'I am going to be more respectful of my professor.' Instead you can identify more specific things that you will be able to clarify what you mean and are doable tasks that will make the goal work." He is out of room on the overhead now, so he goes to the board to list his points. "For example, my desire to be more respectful of my professors, is supported by three behaviors." He repeats them out loud as he writes them.

1) saying good morning to them when I get to class
2) not talking during the class, and
3) not making fun of them outside of class with my friends.

"You will know as soon as you say good morning that at least part of the goal has been fulfilled. And you will know when you have or have not done the others as well."

A girl sitting behind us raises a question, saying, "Mr. Handson, I can say good morning to my professor. But what if I'm not sure I can follow through on the others? In fact what if I'm not sure I want to? Do I make the goal and just do it sometimes, or not make it at all?" I like her question because I know it is easier to make a goal than to keep it.

Scott smiles. "Very good question. Part of what you are asking has to do with commitment. Specific goals make you commit yourself more, but that's the idea of making goals isn't it?

"If we made as a goal only those things that didn't require anything of us, we aren't going to progress are we? Also, if we make a goal with no real intention of keeping it, then when we don't follow through we will simply make an excuse and that will probably be the end of the goal. Is it better to have not made the goal? Probably so. That is what I call low goal integrity. If, however, we have high goal integrity, then when we make a mistake we learn from it, correct it, and determine to really do better. Then we start again, with greater resolve to be successful."

It appears he was ready for this question because he goes back to his stack of overheads, fingers through it and places one on the overhead projector, then continues. "This is my goal integrity spectrum."

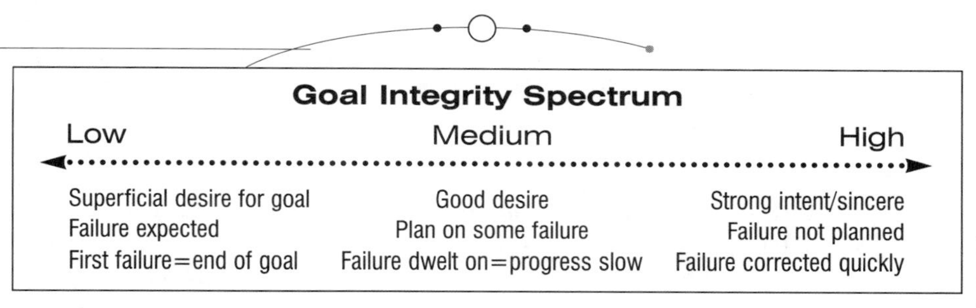

Scott looks at the girl who asked the question. "Is that helpful?" She nods and he continues.

"I'll cover the next concepts fairly quickly because of our time. The next point is, we should have a specific timetable for accomplishment. The reason for this point is similar to the previous two. Specific time frames give us a definite target to shoot for and definite feedback on our progress."

He points back to the goal of two hours of calculus study each day and continues his instruction. "Look at the goal we identified before about studying calculus. If we modified the goal to read 'I will study calculus for two hours each day, from 4:00 to 6:00,' we now know we need to hit the ten hours this week and we would also know where we stand each day. If we miss half an hour one day, we can still make our overall goal by making up time the next day.

"Longer term goals work the same way, and setting a time frame offers the same advantages. We are able to break down the goal into time bites and segments that make it more realistic and manageable."

Scott is at the overhead projector again with his marker ready to continue. "Next, the goals should be written. As others have said before, 'A goal not written is only a wish.'"

He tries to write some more on the overhead even though there's not much room left. I'm having a hard time reading what is on it, but I can make out most of it.

Scott continues, "Writing a goal does a couple of things for you. First, it helps clarify your thinking. When you write a goal, just as when you write anything, you are better able to know exactly what your understanding and position are on a subject. Writing a goal helps you know if you have it clear in your own mind.

"Secondly, writing a goal helps to remind yourself of exactly what it is you are going to accomplish. This is why an unwritten goal is considered only a wish. If you don't take the time and effort to write it down, review it, and remember it, then it must not be very important. In addition to being written, subgoals and tasks must support the main goal identified. If a task is contrary to or unsupportive of the goal it originates from, then there must be an error in logic and alignment."

Scott backs away from the overhead and walks close to the class to make his next point. "For example, if attending too many social gatherings has caused problems with grades, it makes no sense to have as goals the improvement of your GPA to a 3.5 and, at the same time join the campus daters club. The same is true of rewards for accomplishing a goal. It is inconsistent in accomplishing the goal of studying calculus for ten hours one week to reward yourself by not studying at all the next week."

I can tell Scott is feeling a little bit hurried but he seems determined to make sure he gets all the points to us. He must feel it is important that we understand them. The points he has explained so far sure make a lot of sense. I feel excited to try them out.

"The next point is that, if possible, we should share our goals with others who can encourage us. Sharing a goal with someone else increases our commitment to the goal. Having someone to report to will cause us to more seriously consider the consequence of not doing what we said we would do. It also gives us a second source to go to for help and support. If you are struggling with a goal and someone you trust knows about it, they can provide an extra measure of strength."

Scott turns to the class and asks, "Have any of you had any experience with this idea?"

I am surprised that, almost immediately, Troy raises his hand. Scott points to Troy.

"When I was in high school, a friend of mine wanted to stop swearing, or at least cut down. He was struggling with making any progress and came upon the idea of sharing his goal with me by asking me to remind him frequently not to swear. Then he topped it off by promising me a quarter each time I heard him curse."

"Good example," Scott tells him, then asks, "Did it help?"

Troy smiles, "It sure helped me. I made about twenty-five dollars in about a month just by listening to him swear." Everybody laughs and Troy continues. "Actually, it did help. The first 10 or 15 bucks came in the first week. After that the amount decreased, which showed that at least around me, he was swearing less. After about a month he said he couldn't afford it anymore and he called off the deal, but he also felt that the plan had helped him a lot. He said at least he thought more about what he was saying before he said it."

Scott nods his head, "Thanks for the story. It's a very good example of this point." Then, referring to Troy's case, he says, "Setting this specific goal and sharing it with a friend who wanted to help him provided an extra incentive for him to quit swearing. By the way," Scott turns toward Troy again, "why did he ask you to help?"

Troy responds, "I don't swear and we were good friends, so he asked me to help."

"Interesting," Scott says quietly. "It is true that sharing the goal with someone who cares about you, or has insight or experience on the value of what you want to accomplish, is best. One who does not see the value of your goal and cannot do what you hope to accomplish may not offer much moral support or advice to you."

Scott moves back to the overhead projector and points to the next item on the list. "The next thing is that the goal should be attainable but make us stretch. Goals must be realistic, but by the same token we need to have to work hard to get to them. Goals in our comfort zone do very little for us. No stretching means no growth. Let me give you an example." He pulls a paper from his folder and begins reading from it.

"A number of years ago, Motorola chairman Robert Galvin set a goal for the company to improve quality ten fold in five years. Some scoffed and thought it was impossible but they went for it anyway. Many areas were able to reach their goal. It required that they think differently about their processes and how they did things, but that is part of what needed to happen. Sometimes a seemingly unrealistic goal can be achieved if we approach it from a different perspective. Setting high goals usually means changing the way you do things. A reordering of priorities, if you will. But again, that is what goals are for."

Scott again walks toward the class and stops right in front of the first tables. This seems to be his way of putting emphasis on a point. "The setting of the goal and the reaching of the goal are both important, but the real value of the goal is what happens in the process. A goal is to promote and cause change for the better."

Now he walks back to the board and writes down two numbers. "You may set a goal to increase your semester GPA from a 2.77 last semester to a 3.5 this semester. It may be impossible the way you are operating right now. You may have to attend a lot fewer parties, cut out time in other leisure activities, change your sleep habits, and perhaps even cut down the number of hours that you work. Is the 3.5 attainable? Probably. Will it make you reach? Certainly! Is it worth it? That is a decision that you will have to make. It would certainly help your intellectual efforts and probably your job search eventually as well. But the 3.5 in itself does not have nearly as much value as the discipline, habits, and learning you gain in the process of achieving the goal. And just as important is this: what you learn in that kind of effort will cross over into other areas of your life. That will make you a better all around person."

Scott looks at the class and can tell we are getting weary of writing, at least those of us who are taking notes. He smiles and says, "There is one more point I need to tell you, then we will finish up by having you review your own goals. The last point is that you should set times regularly, such as weekly, monthly, or yearly, to review your goals and revise them if necessary. Goals must be reviewed often. The activity of writing them down alone will do you very little good if they are not used. They are to work for you, and for that to happen they must be reviewed regularly and revised at times as well."

Moving back to the board again, he writes the words "review frequency" at the bottom of the list of points and turns back to the class. "The frequency of review will usually align with the time frame of the goal. A daily goal ought to be reviewed at least twice a day. A monthly goal should be reviewed 3 or 4 times a month, and a yearly goal at least two or three times a year. This review and revise process, combined with linking daily to weekly goals, weekly to monthly goals, etcetera, will make the goal setting process work. Too often, both in business and in personal lives, time and effort are spent identifying goals, but then the goals are placed in a drawer and forgotten. The benefits and improvement that could come from the goals will be gone."

Now Scott goes back and sits on the table in the front of the room. He concludes his instruction on the goal-setting process by putting up a summary list of the first four points of the Goal Achievement Process on the overhead projector.

GOAL ACHIEVEMENT PROCESS

Points for successfully setting and achieving goals.

1) We must have personal or individual goals.
   - — We will be and feel more productive.
   - — Some of our goals should be directed to learning and personal improvement.
2) Goals should be specific and measurable.
   - — *Example:* "Do better on my calculus assignments this week" not as good as "This week I will spend two hours a day studying calculus."
   - — Achieving specific goals helps develop discipline.
   - — Easier to know when you have done them.
3) Have a specific timetable for accomplishment.
   - — Better calculus goal is "study calculus from 4:00 to 6:00 every day."
4) Goals should be written. (A goal not written is only a wish)
   - — Clarifies thinking.
   - — Reminds yourself of exactly what you are going to improve or accomplish.
   - — Also subgoals and tasks, and rewards must support the main goal.
   - — *Examples:* — GPA of 3.5 & daters club doesn't work.
     — Can't reward studying one week by not studying the next.

He leaves it on the overhead for a moment and says, "I'll leave this up for those of you who want to copy this in your notes." While we are writing he continues to talk.

"Using the process will lead to achieving goals and thereby bringing us closer to what we want to achieve. Achieving goals requires discipline and persistence. Chronic failure to reach goals can cause lowered self-esteem. If discipline and persistence are problems, it may be well to start your goal setting by making some rea-

GOAL ACHIEVEMENT PROCESS (continued)

5) Share them with others who can encourage us.
   — Increases commitment, and provides support to the goal.
   — Share with someone who cares, has insight or experience.
6) Goals should be attainable but make us stretch.
   — No stretching means no growth.
   — Real value of a goal is what happens in the process as well as the actual result.
   — A goal is to promote and cause change for the better.
      — Example of GPA from a 2.77 to a 3.5.
      — Have to change habits, thoughts and activities but worth it in long run.
7) Set times regularly —weekly, monthly, and yearly — to review and revise goals.
   — Review frequency depends on the time frame of the goal.
      — Daily reviewed twice a day. Monthly reviewed 3 or 4 times a month.

sonable short and medium term goals and review them regularly. This will improve your discipline. Such a goal, when achieved, will return great rewards for the future."

He stands as he makes his last comment. "Remember though, setting goals can provide a direction and motivation otherwise not achieved. A man named Richard Evans was fond of saying that 'If one was not certain as to where he was going, it didn't matter if he got there.'"

With that, Scott pauses for a minute and collects his paper and overhead transparencies. I glance back over my notes before he takes the Goal Achievement Process list off the projector. I have the seven points listed, plus some of the comments from the discussion.

I glance at my watch. There are only about 20 minutes left in the class time. I don't know if Scott is done or has more planned. I hope we are about done, because I'm starting to feel like I am trying to take a drink from a fire hose. Scott answers my questions.

"Now," he's putting his overhead in a packet as he talks, "I want to put you into three groups and finish up by having you work with your goals." He looks at the two girls at the table behind Troy and me. "Would you two join these guys?" He points to us. I'm OK with this, they are both kinda cute.

They smile, and the one who had asked the question earlier responds cheerfully, "No problem." They gather their stuff and make the move. Scott organizes the other eight into two groups as well.

"First, take a minute or two and introduce yourself to others at your table and tell them why you are here." Scott walks back to the front of the room as he gives that instruction.

Troy starts the introductions. He has a way of sounding like he has known you for a long time and that you are already a friend. He tells the girls his name and then tells them I made him come. I begin to protest, but he laughs and then explains that we are both in the same physics class and the teacher suggested this class in our lecture today. His humorous nature seems to set a comfortable tone for everybody. One of the girls, who is named Christy, said she is in the same class and came for the same reason. The other girl is her roommate. Her name is Kellie, and she came along when Christy told her about it.

We chat for a moment about our class, then Scott gives us our last assignment. "I want all of you to now take the goals you made earlier and modify them to fit the criteria I have presented to you. You don't have time to do this with all of them so select one or two, maybe three you can do it with, then do the others later. Then with the last five minutes or so, each of you take a minute and share one of the goals with the rest of the group. This means that one goal will have to be something you don't mind others hearing. OK. Go ahead and start."

We look at each other and I say, "Well let's make our changes and then at five till," I'm looking at the clock, "we'll get back together." They nod and we all turn to our previous goals.

As I look at mine I hear Christy and Kellie whispering something about dates then laugh a little. I don't know if they are talking about Troy and me or a goal they have. I figure either way I might get a date out of this group activity.

I look at my goals and decide to focus on the one year goals I had listed. They look like the easiest to modify to fit Scott's criteria.

I take each one and add some specific, measurable parts to it that I think will suit Scott's requirements. I know I will need to set some shorter term goals with some of these. I mark those I need to review further with a "more work needed" note. I end up with:

Drew Barnes' modified one year goals

1. Pass my classes this year
   — maintain a 3.5 gpa overall
   — not have less than a 3.0 in any class
   — set specific class study goals by the end of the week
   — (more needed)
2. Learn to sky dive
   — attend the open house advertised on the flyer I got
3. Decide on a major
   — attend the engineering and engineering technology open house
   on Thursday
4. Meet some great friends
   — invite Troy, Christy and Kellie over for snacks and a game of Monopoly this weekend
5. Learn some good habits to make me successful
   — attend the classes on setting priorities Scott teaches
   — review all the goals listed earlier by the end of the week

After I have finished I lean back in my chair to look over my newly established goals. I realize that Scott is right about reviewing goals. If I don't check these fairly often, I will likely get busy with other things and forget them. I guess there really is no magic in setting goals, but they can help me keep focused on what I want to accomplish. I still need to work hard. Now I know what I'm working at.

Troy's voice interrupts my thoughts. "Drew, are you finished?"

Troy looks at me but before I can answer Christy chips in, "We are, should we share our goals?"

"Sure, you two first." I want to wait until the last to tell them mine.

"OK," Christy replies, "I'll go first. My goal, at least the one I will tell you, is that I am going to get at least a B+ from the physics class we are in. How's that?"

"Good." I'm curious if there is more though. "Anything else?"

She looks at me rather puzzled and asks, "Do you mean do I have other goals?"

I can see I better be careful what I say. I proceed cautiously, "Oh no, I know you have other goals. We are just supposed to share one. I was wondering if . . .",

"What he means," Kellie looks at me with a smile and interrupts, "at least what I think he means is, do you have other more specific goals to help you make sure you accomplish that goal? You know, things you will do every day or week to work towards the B+."

Troy enters the conversation now. "Let me share mine. I like Christy's goal and mine is similar. I kinda copied one that Scott mentioned. My goal is to study physics for an hour and a half every day. That goal is easy to measure and to know if I have done it. I like Christy's goal too though and I need to add one like that to mine." Christy returns Troy's smile and thanks him. "Grade wise," Troy continues, "I'm shooting for an A, but I have to because that is my major. So I am going to make my goal, 'To achieve an A in physics, I will study physics every day for at least an hour and a half.'"

Christy adds to hers. "I see what you mean now, Drew. So my goal could be, I am going to get at least a B+ in physics by hanging around Troy." She tosses a smile his way and he blushes a bit. "Actually I do know what you mean and I will add more specifics to my goal to make it work." With that she turns to Kellie and asks her, "What's yours?"

Kellie states simply, "Mine is to make my bed every morning and to brush my teeth every night and morning." We all look at her, and Christy particularly has a puzzled look on her face. Kellie explains, "Scott said if we wanted to build confidence to start simple. I'm not too bad at brushing my teeth, but I need to be more orderly in my room." I detect a slight nod of agreement from Christy. "Scott also said that succeeding in one area will help develop discipline in other areas we want to concentrate on. I believe him and intend to test that idea." She is pretty matter-of-fact about it.

I add my voice to hers. "I think it's a great goal. In fact when I am able to find my bed among all the stuff in my room, I am going to make it as well."

They all laugh. I can tell Kellie appreciates the endorsement. She looks at me and says, "Well Drew, we have all shared our goals. What's yours?"

I smile and say "My goal was to listen to all of yours. I did and I'm done." They all laugh and protest, knowing I'm joking. "OK," I give in, "actually one of the original goals I wrote down was to make some good friends here at college. I revised that goal in our little exercise here to read . . ." I pick up my notebook and read more formally now. "I, Drew Barnes will begin my goal to make some good friends by inviting Kellie, Christy and Troy to the dorm this weekend to play games and have banana splits." I put the book down to their laughs and comments of approval. I continue, "So to be more specific, which night works best for all of you?"

Kellie speaks first, "I can't do it Saturday night, how is tomorrow for all of you?" She looks around and everybody seems OK with Friday. "So Friday it is."

I ask, "Is 7:30 OK?" Everybody agrees.

We chat a little and then Scott has every group report that they had done what they were supposed to. He comments that some of the groups seemed to really enjoy the assignment. He is looking at us with a smile when he says it. He then reminds us that he will be teaching a class on setting priorities and other similar subjects then he excuses the class.

As we walk out Troy whispers to me, "Nice move. I mean with the goal. This will be fun. Christy is pretty nice. I'd like to get to know her more."

I smile at him, "I did it all for you Troy."

He sneers at me and says, "Yeah, right."

We walk with Christy and Kellie part way across campus then split up. They live in a different dorm complex than we do. Not much further, Troy heads to his dorm. "See you in class," he says as he splits off. I reply and continue toward my dorm. As I enter my dorm I'm thinking a little about the upcoming date tomorrow night, but more about the things Scott taught us. I'm excited about trying to make my goals more specific, but at the same time have this feeling that it might be harder than I think. For some reason two adages that my dad always said come to mind. The first is "The secret isn't in the saying, the secret is in the doing." He would always tell me this whenever I would come up with a grand new plan for anything. Most of those plans I never seemed to finish. The second was "It's important to endure through the middle because everything seems like a failure in the middle." I also get the feeling that the hardest goals to keep aren't going to be game dates with my friends.

When I get to my room, Stu is gone, bed not made, and stereo still on. I decide I will work on modifying a few more of my goals then do some more homework. One of Scott's comments sticks in my mind. 'We indicate what is most important to us by what we spend our time and money on.' And, it occurs to me, what we spend our time and money on is what we become. These are pretty deep thoughts for a freshman, I think.

Chapter 5

The first week of school is over and went quite well. The date with Christy and Kellie was really fun. We spent most of the evening playing Monopoly. Not a super popular game these days, but it turns out they all liked the game as much as I do. Troy ended up winning, but not without a hard fought battle against Kellie. He finally got the edge when she hit Pacific Avenue and Boardwalk in succession, both of which he owned and both had hotels. Christy and I were pretty much out of the game an hour before they finally ended. At the beginning of the evening I would have guessed that, if we did pair off, I would feel more comfortable with Kellie and Troy with Christy. But the time Christy and I had to talk while they finished the game gave me time to get to know her. A pretty girl, with sandy blond hair, more level headed than I would have first thought a freshman would be. She thinks she's interested in either education or medicine and is taking the physics class to prepare for med school or perhaps to teach science in high school. She hasn't decided yet. She is only about 2 inches shorter than my six feet, and looks like she would be a good basketball player. However, she claims she is too clumsy. The finish of the game gave Troy and Kellie a chance to visit also, and they seem to hit it off as well.

After Troy had driven Kellie into bankruptcy, we all went to the student center for a burger and drink. I tried to talk Troy into paying since he won but we ended up splitting the bill. We then walked our dates home and returned to our own places. We decided we wanted to take them out again, and they seemed to have enjoyed the time with us too.

Now that the first week is over I feel like I'm settled in enough to look for a job. I check the fast food joint across the street from the dorm. They have a couple of openings, but I decide before I take one of them I would check the companies that Dr. Stromberg mentioned. I called home last weekend and talked to Dad. He thought it would be a good idea to get work in a company, but the idea of working in a real company is kind of scary.

I check companies in the order that Dr. Stromberg gave them to me. The electronics firm and the plastic company say thanks but no thanks. As Dr. Stromberg had said they hire co-ops during the summer but don't need anyone now. I'm still feeling the butterflies in my stomach when I get to the front door of Advanced Component Technologies Corporation. I'd hoped the nervousness would be gone by now after talking to the other companies, but not so. I guess because this is a real industrial plant I feel a little intimidated, not like the fast food and summer jobs I've had before.

As I approach the main entrance, I notice a side door about half way down the building that some people are using. They look like employees. There are men and women using the door dressed both in casual clothes as well as suits and ties. Dad said he had a friend who used to work here and there were about 400 employees. He thought it would be a great place for me to get my feet wet in an engineering-type environment.

Dad's friend told him that ACTC makes devices that are used in a number of larger products including airplanes, automobiles, and large industrial machinery. I don't know a lot about it now, but hope I will soon. Even so, I am feeling pretty nervous and wondering if it might not be safer to get a job at the fast food joint by my dorm. I know I can do that. I stand at the front of the building glancing between the front door and the people walking in the employee side entrance door.

My thoughts of copping out are interrupted by a voice beside me. "May I help you?" A well dressed man is looking at me standing in front of ACTC's door, obviously wondering what is wrong with this young kid.

"Oh, I'm sorry. No, . . . well maybe," I stumble, feeling embarrassed about getting caught daydreaming. "I'm here to see a Mr. Walker in engineering. I'd like to see if there is any work here as a part-time intern." I feel good that I got a couple of sentences out. I had gotten Mr. Walker's name from Dad's friend. He said to ask for him."

"You must mean Jim Walker, our engineering manager," the man responds. "I know him. Why don't I take you to his office? Come with me." He starts toward the front door.

As we enter it's clear the lady at the front desk knows him. "Good morning, Mr. Hall." She says cheerfully. I immediately get the sense she is the kind that is cheerful all the time. "Who is your partner?" She looks at me.

"I'm not sure" answers the man I now know is Mr. Hall. "I found him thinking hard about something just outside. He's here to see Jim Walker." He then turns to me. "What is your name?" he asks.

Now, feeling embarrassed, I start stammering again, "Oh, I'm sorry. I'm Drew. Uh, I mean Andrew Barnes. I'm here to see Mr. Walker about a part-time intern job."

Mr. Hall turns back to the lady at the desk. "Jan, this is Andrew Barnes. Would you please fill out a visitor pass for him? Don't worry about calling Jim. I'll take Drew by his office on my way in. It's not out of the way."

While Jan is filling out the pass, Mr. Hall turns back to me. "Are you a student at the University?" He asks.

"Yes sir." I'm determined to be more professional. "I started classes last week. I'm a freshman."

"What are you studying?" he asks.

Though I am still not absolutely sure, I try to look certain. "I'm in engineering. I'm taking some prep classes in math and physics, an intro to engineering and technology class and getting a couple of general ed classes out of the way, too."

He looks at me with a questioning kind of look. The kind you see when you know someone wants to say something but they hold back. I wonder if I have something in my hair.

Jan hands me the pass. "Here you go, Mr. Barnes." She has a pretty smile that is obviously on her face most of the time. "You are now legal for a day. Just don't try to steal any company secrets."

"Oh I won't. I'm . . ." then I see she is teasing me and Mr. Hall starts to laugh.

"Don't worry Jan. I'll keep an eye on him and if I see him with anything I'll call security." He then motions to me, "Come along Drew. Let's get you to Jim's office."

"Who is your intro to engineering teacher?" he asks as we start walking.

"Dr. Stromberg," I tell him.

I see him smile slightly. "Good man," he says. "Listen closely in his classes. He has a great perspective. He may not have as much of a research budget as some professors, but he knows what makes a good person for the engineering and engineering technology fields. He's done some work for us. He is a good worker and a smart man."

As we walk, Mr. Hall points out a few necessary locations — the lunch room, the rest rooms, and a company library and learning facility. I didn't know companies had such a thing as a library.

Mr. Hall stops at a partly opened door at the beginning of a row of offices. He knocks and at the same time says, "Jim, I have a delivery for you."

The man in the office looks up, stands and comes to the door. "Hi Stan," he says, "I wasn't expecting you."

"Well," says my escort, "I found this young man outside looking lost. It seems he's here to see you." I extend my hand as Mr. Walker does the same.

"Hi, I'm Andrew Barnes. I'm here to see you about a part-time engineering intern position." Mr. Walker motions for me to come in then thanks Mr. Hall for bringing me by.

As Mr. Hall turns to leave he looks at me and says, "Drew, if Jim here gives you a job, I'd like you to stop by sometime and talk with me about your classes, if you wouldn't mind. I think I might have some ideas about those general ed classes you want to get done. Jim will show you where my office is when you have time."

I shake his hand, "Yes sir. I'd like to do that," I tell him. Maybe he has a few tricks to get out of some of the general eds. "Thanks for your help, Mr. Hall, I really appreciate it." I'm thinking that if he hadn't caught me outside I might be putting on an apron right now and flipping burgers.

After Stan Hall leaves, Jim Walker, with a wry smile, says, "Already hobnobbing with the high brass huh, Andrew?" My confused look prompts his next question. "Don't you know who Stan is?"

"No" I say, "he walked up when I was standing outside and asked if he could help. Who is he?" I am wondering how dumb I must have looked outside.

"Stan Hall," Jim says matter-of-factly, "is the vice-president of operations here and probably the next president of the company. Real good man, easy to talk to, relates well to all levels of employees. And he's smart. He comes from a technical background but also has good business sense." Then, his look changing to more questioning than statement, Jim asks, "What's this about your general education classes?" Jim's look makes me wonder if I did something wrong again.

"Oh, when we were walking to your office he asked me what classes I was taking. I told him I was in some engineering pre-requisite classes and that I was taking a couple of general eds to get them out of the way. "

Almost as I say it I regret I did, but it is too late. Jim laughs out loud now and a look of understanding covers his face. "You be sure you go see Stan," Jim chuckles. "He sure will have some ideas for you. But Andrew, back to business. Barnes is your last name right?" I nod. He asks, "What can I do for you?"

"Well Mr. Walker, I need to have a part time job. My dad and I thought it might be a good idea to see if there were any openings here so I could get some experience while I work through school."

Jim leans back in his chair. "Well, we just might. I'm glad you came. As a matter of fact we do have a couple of part-time openings. Let's talk about things and see if we can match you up with one of the jobs." I might have gotten lucky here. Let's hope so.

Jim Walker spends about 20 minutes with me. He asks about my goals and what I hope to accomplish at SU. He asks about my interests and the number of hours I can work. He seems satisfied with what I tell him.

He is also interested in my schedule. I explain that I am out of class at noon on Monday, Wednesday and Friday, and 11:00 on Tuesday and Thursday. I expect I could work up to about 20 hours a week but that might be too much. Mr. Walker agrees. He says he would like to hire me but suggests 15 hours a week maximum. He explains that my studies can't suffer because of work and I should concentrate on school.

I'd hoped to work more hours because I need as much money as I can make, but I also realize Mr. Walker is the boss, and his advice is good, so I trust his judgement. We decide I'll work from 2:00 o'clock to 5:00 o'clock every day. That will give me evenings for study and social life and still have a little bit of time during the day to do some homework. It also gives me time to get here on the bus. Fortunately one of the city buses that goes by the dorms also goes by ACTC. It will only be about a 15 minute ride.

With that, Jim takes me to Human Resources to fill out paper work and get an orientation. He leaves me there and I agree to show up at 2:00 p.m. the next day. After completing the paper work and learning a little more general stuff about the company, the HR manager, whose name is Shelly, asks if I would like a short tour of the plant. I've got time, so we head to the floor.

As we approach the floor Shelly reaches in her pocket and retrieves some safety glasses for both of us. I take a pair, but must have a confused look on my face because Shelly answers the question I didn't ask.

"Safety is of highest concern here." She says. "In the shop you will be around some very powerful machines, and there are always things moving or being worked on that could cause harm." Now she points to a large yellow sign at the door going into the shop. It says "Safety glasses must be used beyond this point!" She continues, "Also, you can't wear any open toed shoes in the shop and you cannot wear shorts. Our safety program is one of the best around. We pride ourselves on following proper safety procedures. You'll learn more as you go, and especially as you have a chance to attend our safety meetings. Keep these safety glasses at your desk. You'll need them." She finishes and motions for me to follow her.

Our tour isn't long but it is interesting. She admits she doesn't know many technical specifics about the operation, but she seems pretty sharp to me. The sounds and the activity of the manufacturing floor are exciting to me. The floor looks pretty organized. Though there are a few conveyors, it appears to me that most of the material is moved by carts.

After the floor, we walk back through the engineering offices. Shelly admits she is taking me through the plant backwards of what the product flow probably is, but it was the simplest route from her office. Computer screens and document files are prevalent in the engineering section. A number of offices have sharp looking pictures on the walls that appear to be products the company makes. As we walk by Jim Walker's office, I see a man and woman gathered around a blueprint covering his desk. Jim sees me and calls out, "See you tomorrow," as we walk by.

Shelly then escorts me to the desk at the front door. As I take off my visitor badge, Jan, the nice lady at the desk asks, "Well, Mr. Barnes, are you coming back?" I tell her that I am, impressed that she remembered my name. Shelly tells me that my employee badge will be waiting for me at the front desk when I come the next day. I decide I'm going to use the front door every day. I say goodbye to Shelly and Jan then thank them for their help. Jan's smile could lift anyone's spirits.

I then head out to the bus stop. "This is great!" I think to myself. I'm sure glad that Stan Hall came along when he did. I'm still nervous about what they will expect of me. I feel like I don't know anything. But everyone seems very willing to help and I like the place already. As I see the bus approaching, I feel grateful I've got a job at ACTC.

I smile to myself, and think, "Much better than standing at a register, asking someone 'would you like fries with that?'"

Chapter 6

I'm in one of those reflective moods, thinking about the first few days of school and how lucky I was to get a great job. I know I should be studying instead of sitting on a lounge sofa staring out the window at the trees. I haven't been doing as well in calculus and physics as I had hoped. Quizzes in both classes have proven that point. I know it is still early in the semester, but I've had the problem before of procrastinating, and it catches up on me faster than I think. Dad used to joke about it saying the sooner he got behind, the more time he had to catch up. I know that doesn't really work.

Suddenly I hear my name and I look up to see Russell from my calculus class approaching. He is also in the same ET 101 class with me. We sat together the first day.

"Studying, Drew?" he asks with a smile on face.

I return the smile and reply "Yep. It's a new method called the tree method. You stare at the trees long enough, and eventually you get to the root of any problem."

Catching my pun, he lets out a groan. "Oohhh man. That was really bad." He plops down on the sofa opposite me in a way that my mom never liked. She said it would eventually ruin the furniture. "Hey," I act stern, "Don't plop on the furniture. It will ruin it."

He sees my mocking stern look and laughs. "Oh, I see." He's sitting up straight now, head tilted forward, looking over his glasses. "The repeated impacts of heavy plopping cause stress on joints within the sofa structure eventually fatiguing the assembly and causing a fracture failure thus rendering the object unusable." He stops, smiles and leans back.

"Oh man," I tease, "I'm caught in a nightmare. Please wake me, please wake me!"

He laughs again. "Sorry, I've been studying my statics book and I guess stresses and structures are on my mind."

Russell seems like a great guy. He's taller than I am, wears glasses, and seems very studious. When I first met him last week in class we were to tell something of interest about ourselves. He told us he was a descendant of one of the men who fought with the all-black Civil War company of solders. He's proud of that for good reason. I remember reading about them in high school. They were a very loyal group of men.

We share some light conversation for awhile, when I ask him. "So you're in statics already huh? What's your major?"

"I'm in civil engineering," he replies. "What's your major?"

I look out the window again. "I not sure yet. That's why I'm in ET 101. But I think I'll go into one of the engineering technology programs. Probably mechanical or manufacturing. I'll see which one pays the most."

As I finish he looks at me a little funny. "What do you mean, which one pays the most? All the areas in engineering and technology pay well. Is that the main factor for your decision?" He looks a little surprised.

It seems OK to me. "I want to make a good living. Don't you?" I state the last part as a challenge.

"Well of course! We all want to make a good living, but I want to do something I will enjoy and that will make a difference. I want to go home and contribute as well as have a good living." He emphasizes the part about contribution. He really feels strongly about this.

"Where you from?" I ask.

"Georgia, not far from Atlanta." He replies. "How about you?"

"Not nearly that far away," I tell him. "A place called Springdale about 3 hours away. Close enough to get home for a weekend. That is, if I had a car." I'm curious now. "We're a long way from Georgia. Why here?" I ask.

"Back home there are some real needs in building water storage and drainage systems that will keep the environment clean and functioning well. That's why I came here to study instead of someplace closer to home. The civil engineering program here has a very good reputation for environmental issues, particularly water systems." I can tell he really feels strongly about this. "So I came here to learn from some of the best."

"Is it hard to be this far away?" I ask him.

"A little bit, but hey, it's only been a week. There are differences. The weather is dryer and will get colder than I like. But it'll be just fine." Russell smiles, puts his feet up on the magazine table between the sofas. "So why is money so important to your choice?"

I'm not going to share all my secrets with him, but I can tell he is a good guy and is interested. I have been impressed with his attention in the classes we share and he's nice to everyone. Not really outgoing, just easy to talk to and one of those people you feel you can trust. I put my feet up also. "I just want to make sure I can live OK without having to worry from paycheck to paycheck. My dad's a teacher and it seems our family has never had quite the same things as other people have. We say money isn't everything, but it sure beats not having any. I know jobs in this field pay a lot better than what my dad makes. Don't get me wrong, Dad's a great guy and he is a good father. Like I said before, I figure I will like engineering and technology, but my main reason for getting into it right now is the good salary."

Russell nods a number of times while I am talking. He seems to agree but the contemplative look on his face tells me he is thinking about something else also. He's quiet for a moment then says, "My uncle is in a business that makes a lot of money. He has done pretty well for himself. When I told him what I was going to study, he told me that's what he has always wished he had done. I was surprised. I never knew. He told me he was glad I was doing it right. He explained to me that people in our field do a lot for society and to improve the standard of living." He pauses for a moment, then asks. "Is your dad sorry he is a teacher?"

I realized that I had never heard Dad or Mom say they were sorry. I'm just going on what I thought. Russell and I talk a little more, but I'm thinking about why I am here. I like being around him. He's thoughtful, both in terms of being nice, but also a good thinker. He'll be very good at what he does, and I hope this is one of those friendships that start in college and last for years.

We finish our conversation and Russell leaves, probably to study more. I stay for awhile and continue pondering. What I don't tell Russell is that things aren't that great at home right now. Dad had complications from cancer, and that has taken most of the savings they had in spite of insurance. It's been hard for him to get back to work. I know he and Mom had hoped to help me more at school than they are, but the health problems really hurt them financially. Maybe that's part of the reason the money is on my mind so much. I guess I shouldn't be so focused on that, but it is hard not to. I had hoped to not have to work at school, but I still need to take care of the weekly and monthly things like food, supplies and spending money. I guess I should feel lucky I don't have to work more than 15 hours or so a week. I just hope things work out OK for Dad. His one year checkup after the cancer is coming up in November. Mom says if things look good at that point, then the future looks pretty good. Well, I can't do anything about it right now. Besides, it's time to head to my room for some study, then bed.

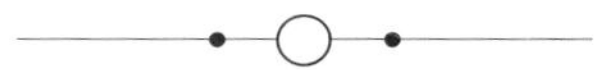

"Come on Stu, turn it off and get to bed, I'm tired." I know I'm sounding more angry than I am, but this is getting a little ridiculous. Just because he doesn't have early classes doesn't mean no one does. He can tell I'm not happy and responds the same way.

"Listen, Drew, this is my room too. If you want to go to bed early that's your problem. I do turn it down, but I'm not going to turn it off. I've got rights too and you can't change that. Don't like it, then leave."

He goes back to the magazine he's reading as if I don't matter. And this is not 'early' as he calls it. It's nearly 1:30 a.m. and I didn't get to bed until after midnight. I've always been a little too quick to anger and this is one of those times I almost lose it completely. "You jerk!" I almost yell at him. "If you've got your rights, then it doesn't matter if anybody gets hurt. No wonder you can't figure out a simple physics problem. You don't have any brain cells that are awake during the day!"

As I grab my pillow and a blanket and head out of the room I look back to see a shocked look on Stu's face. I know he has been having a tough time with his physics class but I shouldn't have said anything about it. I'm tired, I have a quiz in calculus tomorrow, and I'm mad. But, I know I blew it.

After walking around for a few minutes, I find the basement lounge open with an old beat up sofa. I crash there until about 3:30 a.m. then decide to head back to my own bed. The next morning when I'm leaving my room to go to breakfast, then class, Stu turns over in bed and mumbles, "Hey Drew, sorry about last night. I'll do better. You OK?"

He seems to be able to forget these things easier than I can. I tend to hold a grudge. I grunt but don't say anything even though I know I should. As I eat breakfast the whole thing is still on my mind. Russell joins me and says good morning. I just grunt. He asks if I'm OK. I tell him I'll see him in calculus, then get up and leave. Crazy how this all works. I'm more mad at myself than I am at Stu. Even though it is a beautiful fall day, I still let this eat on me. I know I've got to change somehow; it's just hard to know how sometimes. As I walk to class, I decide for now I've got to forget it and concentrate on the quiz I have to take. I resolve to apologize to Stu soon. That at least helps me a little. By this time I'm at my calculus room and the professor is getting ready to start.

First we have the quiz and it takes half the class period, which is longer than I thought it would take. I didn't do very well. I'm going to have to spend more time on this class if I want to pass. Then the professor starts. "Today," he begins, "we are going to talk about how Archimedes discovered a more accurate approximation of pi and how that relates to calculus."

I force myself to take out some paper and a pencil. I know I will need this sometime so I better take some notes.

Continuing, the professor writes the word "limits" on the board. "Calculus is the first math discipline where we talk about limits. Limits can also be considered the main focus of calculus. You may even think of calculus as trigonometry and algebra with limits. Let's introduce the idea with some history."

He puts an overhead up that is subtitled, "Achimedes' Measurement of the Circle." He then continues, "Archimedes wanted to define a more accurate approximation of pi." Pointing to the first figure on the overhead he explains, "He first drew a circle and then circumscribed it with a polygon, and inscribed it with another." Using his pointer, he traced the two polygons on the overhead, "Thus ending up with one on the outside and one on the inside of the circle. Then he measured the perimeter of the two ploygons, in order to give two approximations of the circumference of the circle. One approximation would be less and one would be more than the actual circle. He then calculated the perimeter of the two polygons using formulas from plane geometry."

Now he points to the other figure and continues his explanation. "He then increased the number of sides of the polygons until he felt that the errors between the circumference of the circle and the perimeters of the polygons was very small. At that point, he had used a 96-sided polygon. Remember he was doing this without the aid of computer drafting and design systems. Quite an achievement by itself. Archimedes ended up with a number that showed pi as greater than 3 10/70 and less than 3 10/71. These best estimates of the average of the two bounded polygons yielded 3.14185 or a error of about 0.0002 from actual pi."

He stops for a moment and moves away from the board. "Not a bad job, don't you think? The point of this discussion is not the value of pi but to illustrate the concept of limits, as you see here in Archimedes' work. The secret to getting a close approximation to pi was to keep reducing the error between the circumference of the circle and the perimeter of the polygons. The limit would be taking that error to an infinitesimally small amount. That is a fundamental part of calculus. Using this idea of limits we can then derive, a common term in calculus, areas and other values for many non-standard shapes."

I think now I can see the concept of limits even though I know there will be a lot of problems that will take a lot of time. Some of my pre-cal from high school makes its way out of the cobwebs in my head as well. I check my notes of the things the professor has written on the board, then try to keep pace with his discussion. Now he is moving on, talking about limits of functions and giving some examples. I'm still not sure where I will use this, but I have a gut feeling it will show up somewhere. I determine to take good notes for that purpose and am more successful now at keeping my run-in with Stu out of my head.

I didn't see Stu at all before I left for work, but that's not surprising. I don't want to have to face him. Maybe he should have been more thoughtful, but I was pretty bad myself. After I get home from work, I eat and then go to the library to study. I hope to avoid seeing Stu. About midnight, I decide it's time to head to my room. I listen at the door for a minute and don't hear anything. Even though the light is on, Stu must not be home. He always has his music on, but I don't hear anything. I open the door and go in. I'm surprised to see that he is there. He is sitting on his bed reading his magazine with a new set of earphones on. He glances up at me and says, " Hi Drew, how ya doing?" He acts if nothing happened. I mumble a hello and put my stuff down. Now I feel worse. He has bought a set of headphones, obviously to make it easier on me. I admire his ability to not hold a grudge. I wish I could do that. I move around getting ready for bed and Stu just sits there listening to his music and reading. Finally, I climb into bed. I notice Stu get out of bed and turn out the light. It's then that I notice a desk lamp he also bought that he can use to read without having the main room light on. I look at him as he puts his headphones back on. He gives me the thumbs up sign. I finally smile and return the thumbs up but don't say anything. I decide I will apologize tomorrow even though I should do it now. I turn over and smile to myself. Maybe he's not perfect, but he is a pretty good roommate. Unfortunately, my smile disappears when my calculus quiz score comes back to mind. It reminds me I need to adjust my study schedule for at least awhile. I'll do that tomorrow too. Then at the faint but certainly bearable sounds of Stu's music, I fall asleep.

Chapter 7

Now I know right where I am going as compared to my first visit to ACTC. As I walk through the parking lot at ACTC from the bus stop, I see some reserved parking spots with names and positions on signs at the head of each spot. I decide I'll walk this way every day and use the time to memorize the names of the people in company's management. I notice the one in front of me, V.P. Sales, Barbara Oldmer. The next one is familiar: V.P. Operations, Stanley Hall. "Boy", I think to myself, "I sure am glad I didn't know that when I first met him. I would have made even a bigger fool of myself than I did." I notice the vehicle in Hall's parking spot is a 4X4 Chevy Pickup. New pickup and a nice style. Interesting that Mr. Hall drives a pickup to work. I'm still thinking about my chance encounter with Stan Hall when I reach the front doors.

When I walk in Jan smiles at me and says, "Good afternoon Mr. Barnes. How are you today?" She seems genuinely interested.

"I'm fine, Jan. Why don't you call me Drew? I'm not old enough to called Mr. anything."

She laughs and says, "Whatever you say, Mr. Drew. Here is your employee badge just as we promised. Do you know where you are going or shall I call Mr. Walker to come and get you?"

I tell her not to bother. I am sure I remember how to get to Jim Walker's office. She wishes me well as I leave the reception area.

As I walk to Mr. Walker's office, I think about how Jan treats people. I know ACTC is doing very well, and I have to wonder if one reason isn't Jan at the front reception desk. She could make anyone feel glad to be here.

ACTC has been a pretty successful company. I happened to mention to Dr. Stromberg after class this morning that I had gotten a job there. He seemed a little surprised but commented that it is a good company. He told me to pay close attention and he was sure I would learn a lot. He said that ACTC was well managed, had a good handle on market needs, and treated customers well. He also said they had sound design and manufacturing practices. I could tell that he admired the company and I could also tell that there was a lot of meaning behind the word "sound practice" that I did not understand. It made me feel very lucky, and proud, to be working for ACTC. I really can learn a lot working here.

When I get to Jim's office I see he is in a heavy discussion. It looks like the same man and woman who were in his office yesterday when Shelly was taking me around. I stand for a minute not wanting to interrupt them. I hear the woman say, "Decreasing the clearance between the body wall and the board concerns me. It will restrict airflow that we need for cooling, and if we get too close I am afraid vibrations will cause the wall to deflect too much. It might cause damage to the components."

Just then Jim looks up and sees me at the door. He glances at his watch and says, "Good, 10 minutes early, I like that." I think Jim said it in part for me to know that being on-time is a good habit to have. I make a mental note that I must not be late. "Come in," he continues, "I want you to meet a couple of our engineers."

Pointing to the woman he says, "This is Debbie Anderson. She works on the electrical side of the house." Then to the man, "Carl Sanders is in our mechanical area." I nod at them both, shake their hands and say hello. Looking at me now, Jim says, "This is Drew Barnes. He is a student at SU and will be working part-time for ACTC 15 hours a week. He starts today."

"Where are you going to be working?" Carl is looking at me, but before I can say anything, Jim interrupts

"I first thought I would have him helping in mechanical. But this morning I was talking to Amy Olson and she thought it might be good to have him with Vic Hollins, our design-manufacturing liaison. Vic could use the help and . . ."

"And" Carl interrupts, "Drew could get a good exposure to a number of areas in the company if he is working with Vic."

Jim and Debbie both nod. "Good idea," Debbie adds, "and besides, Vic would enjoy teaching a new engineer the ropes and he knows the company well."

They all nod and then Jim turns to Debbie and Carl. "Why don't you two keep working on this? I'll find Amy and introduce Drew to her."

As we leave his office, Jim says, "I hope you like working in Amy's group. She manages the manufacturing engineering group. Vic serves as a liaison between the manufacturing group and the design people. You will get to know a number of functions and people that way." Then, as if to read the concerns that were in my mind, he added. "We will still get to see each other and work together some. We work a lot with the manufacturing engineers, and we see Vic the most."

I think out loud, "I hope I can handle whatever I am supposed to do." As we continue to walk Jim assures me that I can.

"What we want most are people who work hard, are willing to learn, and do their best to get along with and support others." I take a mental note. I think I just picked up three more things that I can learn from Jim. I decide that I need to pay close attention to how people act and work and not just the tasks I am assigned. It seems to me there is more to being successful in this company than just being able to solve engineering problems.

We soon come to an office located just off the hallway that leads to the manufacturing area. Jim introduces me to Amy. She's a well-dressed woman who looks to be in her mid-thirties. He explains that I am a new part time employee and that he thought it might be a good experience to have me work in her group with their liaison functions.

She speaks quietly but seems confident enough. "I think that would be good Jim," she tells him. "Vic is good with people and well organized. He will be able to keep Drew busy with things that will benefit our group." Then she turns to look at me and extends her hand. "Welcome to the group, Drew." She smiles as she shakes my hand. "We happen to have an empty desk right next to Vic's office. Let's get you situated there."

Jim turns to me and wishes me luck. As he is about to leave Amy's office, he stops and turns. "By the way Amy, I've just been talking with Debbie and Carl about the design of the new S-line device we are working on. We need to get someone from your group to help us out in the producibility aspect of the product. We are meeting again tomorrow at 2:30. Do you have someone who could be there?"

Amy nods and says, "Someone will be there. We need to know what the design people are up to so we can be ready to pilot the new device." Jim thanks her and leaves.

After getting a couple of 3 ring binders from her office, she takes me over to an empty desk. "Here Drew, you can use this desk. There are notebooks in the cabinet by Sharon, the secretary for the group. A copy machine is there as well. Pencils, pens, staplers, and so on are there, too. Most of the group are in meetings, where we spend a lot of time, so I will introduce you to them later. Vic will be back in an hour or so. In the meantime, I would like you to review the kind of products we deal with here." She hands me the binders she brought from her office. "Especially take a look at the sections that describe what solenoids and actuators are and how they work. If you still have time before I am back, then look at the S-line products to become familiar with them. You will probably work with that line the most." She pauses for a moment then, as she leaves, stops and smiles again. "Drew, it is good to have you here. We need your help and I think you will like it here also." With that she leaves me standing by the desk. I begin to feel scared all over again. I wonder what am I doing here. I'm just an 18 year old kid starting out in school and not sure what I want to do with my life. Now I'm standing by a desk in a successful company with experienced engineers and managers paying attention to me and telling me they are glad I am here to help them. It seems like a dream, and I figure I will wake up from this any minute. I decide that until I do wake up, I better figure out what a solenoid is, how it works, and what it does.

A little over an hour later, I am reading about the ACTC line of products referred to as S-line. I hear someone go into the office that Amy said was Vic's. After a few seconds he comes out and walks up to my desk. "Hi," he said extending his hand toward me, "I'm Vic. I saw Amy on my way back and she told me you would be here. I hear you are going to teach me a thing or two."

Vic is a strongly built man no taller than I am but huskier. He's not really fat, in fact he looks like he could have played linebacker in college. He has a pleasant natural smile that immediately sets me at ease. A little bit of balding on top makes me guess he is about 40, give or take a couple of years.

I blush a little at his statement and quickly respond, "No, no, I am just trying to learn about what ACTC does from these books." I point to the binders on my desk. "I really don't know much, but I can learn."

He looks at the binders I have been reviewing then picks up the one that explains what a solenoid is. "So," he says looking up from the binder, "do you know what a solenoid is?"

I know this isn't a test but I still feel a little nervous trying to answer the question. "Well, I have heard of them before. I know my dad has mentioned a solenoid when he has worked on his truck, and I've heard them mentioned in other places. From the book, I can see that they are some kind of electrical switch, but I don't quite see how they work." My nervousness disappears a little when he smiles and tosses the book back on the desk.

"Not too bad for the way that thing is written." He seems a little frustrated about the documentation in the binders but continues. "Do you know what an electro-magnet is?" He looks at me and I nod yes. "Good. Plain and simple, that is what a solenoid is." He can tell from my look that I am little bewildered. "Look here." He takes a piece of paper and sketches a simple diagram and explains.

"Electricity going through a wire creates a magnetic field, you know that from elementary physics, right?" I nod yes again. He continues, "If I make a coil, then I can increase the strength of the field if I wrap it around a piece of iron or steel. This piece of steel is hollow, or at least made such that it has an opening on one end. Then I put another piece of iron in the end that can move."

I interrupt him. "That's the plunger, right?"

He smiles again. "That's right, well done. With this piece, the plunger, which can move in and out, I have a solenoid. Put a current through the coil and the plunger moves in, or it could move out depending on how you build the solenoid. Stop the current and a spring at the end pushes the plunger back. Therefore, you see that a solenoid is basically a spring-loaded electro-magnet. It is a switch in a way because it does turn things on and off as it moves back and forth."

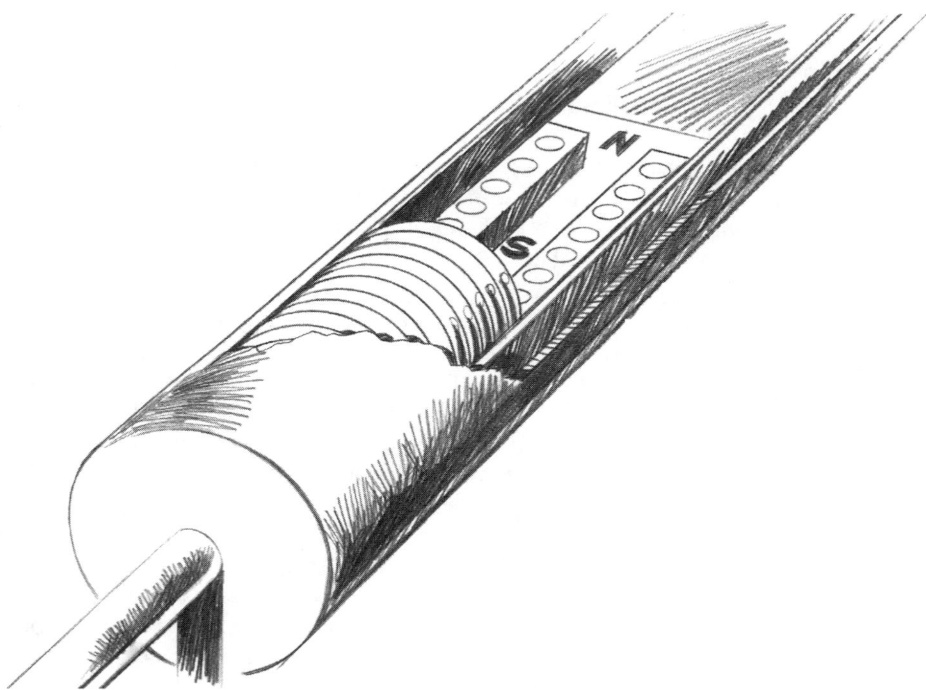

I am impressed that he made it sound so simple. It makes sense to me but I know there is a lot more to a real solenoid. At least I understand the basic operation.

"Wow, that's great. You make it a lot more understandable than the book." I realize I better be careful that I don't rip too hard on the book. "What I mean is your explanation is easy to understand. It's not that the book is bad."

He laughs, "The description in the book worries too much about volts, amps, reducing chatter, heat, and other technical jargon. I'm afraid we are trying harder to impress people with our fancy language than we are trying to help them understand what a solenoid is. Of course, most who deal with us are very familiar with what a solenoid is and don't need an explanation of the basics. Anyway, there are a lot of details that are more than what I have described, but these are the fundamentals." He picks up the other binder. "Looks like you have also been looking at our S-line product. Good. I have some work for you to do, if you are up to it already."

I hope he doesn't mean I need to design something already but I don't ask any questions. I just tell him I'd be happy to do whatever he needs. He goes to his office and comes back with a small stack of papers.

"These," he drops the papers on my desk, "are failure and warranty reports on the S-line products. I would like to have you go through them and count up the number and types of failures on these parts. I have a hunch about them that we have never looked at, and I think we ought to investigate. I just haven't had time to do so. How many hours a day are you working?"

"Three right now, but maybe I could sometimes work more if you need it."

He shakes his head. "Nope, you're in school and that's your first priority. If we need a day or two of 4 or 5 hours maybe we can work it out. You go ahead and plan on your three a day. Look through the binder on the S-line and get familiar with that. That is the line you will be doing the failure count on. You might also want to go back and review the descriptions of solenoids and actuators. It is good to be very familiar with the products. You will probably be interrupted with a meeting or two. We can't go a day without having a meeting. It will probably take you about a week to get this information together. That will be fine. I've got other things to finish anyway. Come here a minute." He leads me to his office.

His office is not too large but comfortable. Other than his desk he has five four-drawer file cabinets, all with computer-printed labels on the drawers. I recognize the S-line label on two of the drawers. His office is really quite neat. On his desk are a number of piles of papers, a phone, and a computer display. There are pictures of his family on his desk and some of what I assume to be his kids on the walls, as well as some pictures of ACTC products. I also see a 10 year service certificate and next to a picture of what appears to be Vic receiving an award from someone. I would also guess, from the pictures and a couple of models, that he is interested in aviation. One of the pictures on the wall is him beside a single engine plane. I'm wondering if he has a pilot's license when he begins speaking.

"When you get to the point of needing to put the information in a chart or table, you can use this computer. The spreadsheet package will do some simple charts. It has a statistical package on it too, but that takes awhile to learn." He stops for a moment and looks at me. "Do you know how to use word processing software?" I tell him I do, but that I haven't used the chart stuff much. He doesn't seem worried. "Most important, just get me a count."

He sits down with me and shows me what to look at on the form and what information to count. It doesn't look too hard, at least counting the failures. There is a lot of other information on the form as well, but he doesn't explain it all. He really does just want the number and type of failures. He tells me that he thinks springs are failing in some of the solenoids much sooner than they should and wants to see if it's true.

After a few more instructions and some general talk I head back to my desk. I feel both excited and a little nervous. My first task doesn't seem too hard but seems to be something that Vic is really interested in. It looks like a good way to get to know something about the product and the company. By the time I get back to my desk, I only have a half hour till time to quit. I decide I will read through the S-line binder a little more and then leave. I'll start the count on the failures tomorrow.

A couple of minutes after five as I am getting close to the end of a section, I hear Vic's voice. "Heading out soon, Drew?" I tell him yes. "Come on, I'll drop you off a couple of blocks from campus. It's on my way to an appointment I have in a few minutes." I accept and wrap things up.

On the way out I ask Vic what kind of an background he has. "I came out of an industrial engineering management program. It is one where you learn fundamentals of engineering but not as much detail as straight engineering programs. Instead, I took more classes in business and management."

As we cross the VIP parking lot I hear a somewhat familiar voice behind us. "Vic, is that Mr. Barnes with you?" We both turn around and see Stan Hall.

Vic answers him. "Yep, it is. You know him already?"

"Yes, we met his first day here. Have we got him busy yet?"

"Sure do," Vic responds, "I'm going to have him help me with a failure analysis on the S-line series."

Stan Hall chuckles. "You really think there might be something there don't you? Well, I hope you are right. We never have taken a serious look at why sometimes they are so good and other times they are not. I would love to see us do a better job of keeping the customer happy. Some customers are not very satisfied with the performance of the S-line. It looks like a great project for Drew to help on. Good luck."

We have reached Stan Hall's truck in the VIP lot and we say goodbye. Vic turns to me and teases, "Making contact with one of the company heads on the first day! Smart move Drew." I try to explain it was all an accident but don't tell him the full story. Vic teases me a little more but before I can respond, which I think better of anyway, we are at his car.

I go back to the question I was asking Vic before we saw Mr. Hall. "Why did you choose a technical field to study and work in?"

Vic ponders for a few seconds then responds. "Well first, asking the questions is a good idea. You ought to take the chance to ask a number of technical people at ACTC. Not being from a traditional engineering program, I will not be able to give you the same perspective others will. But you asked my reasoning." He takes a minute more to think as we pull out of the parking lot. "I have always enjoyed knowing how things work and being able to fix things. I still do most of my own car and home repair work. I guess that means I have a natural inclination toward this kind of work. I suppose many of the things you hear about engineering types are true of me as well as of others at ACTC. I enjoy solving a problem or figuring out a better way. Just like this analysis of the S-line solenoids you're helping with. I have a hunch there is something that can be made better. I want to find it. I also enjoy the management side of business, especially coordinating the work of other technical people."

As Vic talks I can relate to the things he is saying. I feel pretty confident that I do want to be some kind of engineer. I have many of the same kinds of interests Vic is describing. I still have a little of a nagging fear that maybe there is something else, something that would be better for me. As if he has read my mind, Vic speaks again.

"You know Drew, I know you feel you need to make a decision about your career soon, and you should as soon as you can. Deciding on a major and working toward that goal will make your academic endeavors more successful and interesting. Either right now or at some time in the near future you will be concerned about making the wrong decision. Relax about that. Your interests and talents have brought you to where you are and you need to trust them. It is a good and rewarding life to be working in a technically-based company. Take the next few weeks, work hard in school so you don't just learn well, but develop good habits, and learn all you can at ACTC. You really can't change much until at least the semester break, so take advantage of the opportunities you have. Observe people and things. You'll get your questions answered."

"Well, in this kind of field you make good money too, right?" I wonder if I should have asked.

He thinks a minute and responds. "Yes the money is not bad, but that's not the only reason to go into the area. In fact, it's not the best. You better enjoy coming to work, too. Listen Drew, just take your time and observe and learn this semester. Both in school and at work. Things will start to become pretty clear for you."

I think he is right. I've got some time, and I do feel pretty good about what has happened recently. I want to ask a few more questions, but we are coming up on the intersection where I will get out. Vic speaks again. "Let me give you a piece of advice, Drew. How you handle yourself in school, not just grades, but the habits you develop, will in large part dictate how you handle your life. Use this chance to develop yourself well. Enjoy college and the good things that come along with it, but don't go overboard." He pulls over to the curb and lets me out. "See you tomorrow, Drew," he calls as I close the door. I wave and he pulls away.

I have a three block walk and find myself wondering if I just talked with a coworker or my dad. I know he gave some good advice, but I still feel impatient about deciding on a major. I know the advisement people at school will be pushing me to declare something. Vic is right, after this semester, between school and work, I will have a better idea of where to head.

The room is empty when I get there. Stu must be out with friends again. I drop my backpack on my bed and head to the cafeteria to eat. I'll have some homework to get to soon as well. As I leave the main door, I see Russell. He's going the same way and comes up beside me. "Hey Drew, how are you doing? You seemed a little down the other day. How about after you eat we play some ping pong and then head out to a movie?" I don't know if it's Vic's advice or just lack of interest in a movie.

"A couple of games of ping pong sound good Russell, but I think I'll skip the movie. I've got a lot of homework."

"Ahhh," Russell tells me. "One of those kind huh, Mr. Study."

"Not really," I tell him, "but I'm trying. Oh, and sorry about being grumpy the other day." Russell just shrugs his shoulders and off we go.

So after dinner and the best-out-of-five ping pong match, it's back to my room and hitting the books. It's been a good day, and I like where I am right now. I feel better since apologizing to Russell, but I'm glad he doesn't say anything more about it.

About an hour later Stu comes in. I remember my commitment from last night and apologize to him. He takes it well, just as I thought he would, and kindly says that it was partly his fault. I smile, knowing even if it was, I could have handled it a lot differently. There was no need to wait until I flew off the handle. Anyway, it's done and I have learned, or at least I hope I have learned from it.

Stu asks, "What are you studying? Physics or something?"

"No, actually it's history. Hey Stu, listen to this." I read from the book to him, "'Our time is the most advanced the world has ever seen with respect to the products, services, technology and conveniences available to us. Because we live in a time when technology is moving at a pace unprecedented in history and have more knowledge available literally at our finger tips than ever before, there can be a tendency to view past civilizations and societies as less intelligent than ours. It is not true, however, that we are a smarter or more thoughtful people than have ever lived. History proves that deep-thinking, innovative, and creative people have always existed. We are fortunate to be living at a time, place, and in a situation that places us in the very middle, so to speak, of the greatest technological age in history. We have more opportunity for education and learning than ever before. This has come about in large part because of the discovery, and increased understanding, of scientific principles over many centuries. Throughout history the activities now commonly described as engineering and technology have had as significant an impact on the standard of living of civilizations as any other single activity or function.'"

I look at him, and he is still lying in the bed, looking half asleep.

"Did you hear that Stu? 'As significant an impact on civilizations as any other single activity or function.' Whadda ya think of that?" I know I sound proud. "This engineering and technology stuff is for real."

Stu doesn't seem bothered at all. "No question." He answers quickly. "You engineering and technology types make the world go around. All I want to do is take your great ideas and sell them. You don't want to, and somebody needs to. Let it be me. That's what I say."

He is right. I don't really care to do the sales stuff, and it is great to have people like him who like good ideas and want to make them available to others. And we make darn good money too. That's what I want right now.

"What are you reading anyway? Your engineering book?" Stu is sitting up on his bed now looking at me.

I hold the book up so he can see the title. "I don't have it yet. This is my History of Civilization book. It has this neat little section on the history of technology and stuff about the effects of technology on history and the development of civilizations. Here's an example, listen to this."

He leans back on his bed and I turn my chair to face him, put my feet up on my bed and read. "'James Burke describes the snowball effect of some key inventions he calls "triggers." These key inventions, according to Burke, cause major changes in the state of the society. A real paradigm shift occurs in that no longer is the old pattern the governing one or the previous method the most effective one.

The new discovery changes how people think, work, live, and what they have or can acquire.'"

I stop reading for a minute and ask Stu. "Guess what would be the first major trigger? What invention do you think would have a significant impact on early society, say four or five thousand years ago?" He looks at the ceiling and twiddles his thumbs as if he was pondering. "I don't know." He pauses, "Huummm, maybe the wheel?" I'm not sure if he is serious or joking. Just as I am about to tell him he suddenly sits up. "NO, I'll bet it was a material, you know steel or iron or something."

"Close." I'm surprised. He seems interested in this. "Listen to this." I start reading from the book again.

"'According to James Burke the first "trigger" for technological advancement as far as recorded history was the plow.'"

Stu looks at me in disbelief. "The plow? You've gotta be kidding."

I shake my head. "Nope. Hold on. Let me finish." I continue reading.

"'This ingenious device allowed people to settle towns and cities instead of having to roam for food and shelter. By about 3000 BC the people cultivating the alluvial plains of the Nile. . .'"

"What in the world is alluvial?" Stu interrupts me again.

I answer him. "It's like the delta of a river. You know, the area made by the deposit of dirt and such from the river Nile. Now let me finish." I start the sentence over.

"'By about 3000 BC the people cultivating the alluvial plains of the Nile, the Tigris-Euphrates, and Indus valleys were producing so much food that laborers could be hired to make tools, dig canals, and provide other functions. Settling towns and cities allowed them to develop other trades and thus provided the means for them to increase their holdings, and as a result, improve their standard of living.'"

I stop reading and close the book holding the place with my finger. "So this guy Burke says that the invention of the plow was the key thing to the development of society for that time. Before that, people had to chase food and water just to stay alive. After the plow, some people could stay put and raise food on farms while others could do other things. Neat, don't you think?" Stu is looking at me with that pondering look again. I haven't yet figured out if it means he is thinking, brain dead, or confused. Finally he speaks.

"So, what you're saying is that you and I can come to school and learn stuff we want to, partly because we don't have to stay at home and help on the family farm like people 200 years ago in early America had to?"

I guess the look was a pondering one. I hadn't thought about the application to our day. "Well I don't know, but I guess that's kind of the same thing." He is pretty thoughtful about this.

I continue. "I know 200 years ago most people had to farm in order to live. I guess it is kinda the same thing. The advancement of tools and technology in agriculture has allowed the tables to be turned. In fact, less than 10 percent of the population works in agriculture because the advancement of tools and technology has made it possible for less people to do a lot more. And just as the book says," I hold up the book to make the point. "When people can worry less about food and shelter, then they have time work on other things and the standard of living increases."

Stu speaks up again. "And you're saying that engineering and technology is responsible for all of this progress?" He's acting a little defiant now so I back off.

"No, not all of it. But what I am saying, or I guess what I think the book is saying, is that the kind of work that engineers and technologists do is critical to the advancement of societies. If you don't have people who design and make things, how are you going to advance the standard of living for a society?" I think I make a good point, but Stu still isn't completely sold.

"What about all the damage this stuff causes to the environment?" Stu is taking the defensive again. "Maybe it all needs to stop." I don't know if Stu is genuinely concerned about it or just looking for an argument. I decide not to push it too hard because in some respects at least, he probably has a good point. I do want to challenge it a little though.

I continue my argument. "My dad thought it was funny that even the people who sometimes complain about industry and technology use the very things made by the companies that they fight. Computers, cars, gasoline, telephones, all kinds of things that are designed and made are tools that the people use to fight against the companies who make them. Maybe there are some problems, but even those problems are almost always best solved by advancements in technology, not by ignoring or destroying it. True, we have some problems. But shutting everything down won't solve them."

Stu has that pondering look again. Maybe I came on too strong. "Yeah, I see your point. Well, I still can't get excited about taking the physics and stuff to do that. I would rather sell the stuff you make. You're right though, what you guys do is important."

Stu lies back down. I think he's had enough. I read a little further. The book makes the point Stu and I were just discussing. That is, that technological advancement should be wisely managed by the companies and industries that work in the area.

This is an interesting and an encouraging view for me. I know engineers and technologists are often seen as nerds, not as strong contributors or leaders of society. It seems pretty obvious to me that they and their work are very important to society. I know other activities and areas of study are important also, but this gives me a sense of pride about what I think I would like to do. I'm interested in this area because the money is good, but it is obvious that I can be a good, contributing member of society as an engineer or technologist and that there is no reason that I can't be a good citizen and maybe even a leader as well.

I lie down on my bed now, realizing I might just drop off to sleep. But before I do, I feel a sense of satisfaction about a career in engineering or technology. I chuckle to myself. Kind of funny, I think to myself, that I got this from a history book. And I feel a lot better having apologized to Stu and talked with him for awhile. He's a good guy. Different interests and talents than me, but I learn a lot from him.

Chapter 8

As I sit in the Intro to Engineering class listening to Dr. Stromberg talk about making decisions based on good data, I can't help but think about work. I have been working on the S-line failure project for about 3 days now, and will soon have all the numbers recorded. Dr. Stromberg's lecture today is introducing ways to look at sets of data and the importance of having good data to make decisions. This is just in time for my project at work.

Stromberg is talking again, "It is critical in this step," I wonder briefly what step he is talking about. I must have missed it when I was thinking about my project at work. I had better pay attention or I'll miss more. . . . "to let the data tell you the results and not bias the data by your own opinions or expectations. It is also here that, in terms of the hypothesis, you either reject, modify, or accept the hypothesis and proceed from one of those decisions to further work. An example you may remember from your earlier educational experience is that of plants and water. Your basic assumption may be that plants need water to live and from that develop a hypothesis that a certain plant, not watered, will wilt away. You then develop a careful procedure of testing two of the same kind of plants in the same conditions, only one watered and the other not watered. After very carefully following the procedure for a specified amount of time, from the data you gather, you can conclude if the plants actually do need water or not. Based on what you learn from this experiment you may choose to test other plants further or, in different conditions, to test other aspects of plant survivability."

Stromberg now goes to the overhead projector and puts up a slide. He continues, "The following statement by Doyle emphasizes the importance of letting the data and the results of observation and fact direct your conclusions and further efforts. Too often this does not happen."

> *It is a capital mistake to theorize before one has the data. Insensibly, one begins to twist facts to suit theories, instead of theories to suit facts.*
>
> Doyle

Getting another slide from his pile Dr. Stromberg then adds. "Mark Twain said it another way as only Twain could. But it means pretty much the same thing."

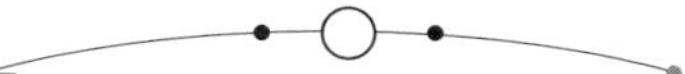

First get your facts; then you can distort them at your leisure.

Mark Twain

Professor Stromberg continues "To be most effective in your work, let the facts, the data, and the experiments stand for themselves. If they can't stand for themselves, then don't make them what they are not. Rather, if more work is needed, then do it. If the experiment or the data is invalid, then don't use it. As is noted by Doyle, we too often want to make the data or experimental results fit our version of things. Resist that urge. Even in the scientific method, when a scientist generates a hypothesis and then conducts experiments, he must let the data speak for itself. If the data does not prove out the hypothesis, then more work must be done." He stops for a moment, then adds. "I hope you realize that this quote by Twain is in jest. You should not distort the facts, of course. Twain is not telling us how to work with facts, only commenting on what too often happens."

Even though I am not really conducting an experiment, I know the counsel he is giving us will apply to my work. I know Vic has pretty strong opinions about the causes of failure for the S-line solenoids and I am sure others do too. I have even developed some opinions from going through the failure sheets. I have to be careful not to let these opinions skew the real results. But I am wondering how to best present the data. I know it is important to make it clear and simple. Amy had stopped in to see how I was doing yesterday, and when I showed her what I was up to she told me to make sure I presented the data in a clear and straightforward way. She said that even though nearly everyone there has a technical background, it is still best to use simple methods to communicate the information. Maybe I can get an idea how to do that from Dr. Stromberg. I raise my hand and Dr. Stromberg calls on me.

"Dr. Stromberg, what is the best way to present data so it is clear and easy to understand?"

He takes a step or two in my direction and stops. "It really depends on what kind of data you have and the context of the data. That is, what form is it in, what form has it taken, what kind of an experiment you were doing, and with whom you are sharing it. Do you have a specific example for us that would clarify your questions?"

I am hesitant to say too much, but continue, "Yeah, kinda. I work at a local company that makes solenoids. I'm not doing any experiments, but right now they have me working on a project where I am checking failure reports to see why some solenoids fail. I have been wondering what is the best way to present the information I get." He has not taken his eyes off of me and seems very interested.

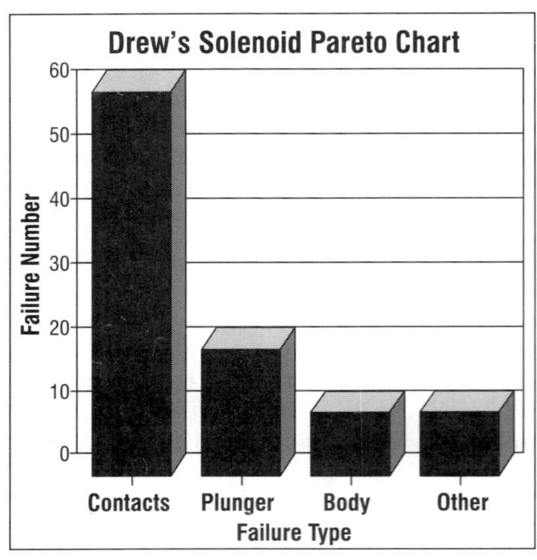

"It is ACTC?" he asks. I nod. "In that kind of situation there are a few chart types to look at but one you will want to give serious consideration to is the Pareto chart." He moves back to the board now and begins to draw a chart. He continues talking as he draws. "The Pareto chart is a simple way of showing, in your case, the number and type of defects organized in order from the largest to the smallest. Basically a Pareto chart is a bar graph ordered, in this case, according to the number of failures."

He points to his figure on the board. "For example, perhaps you have 100 solenoids and 60 of them fail due to poor contacts." He labels the first column accordingly. "Then you have 20 due to plunger failure," he labels the second column, "10 to body failure and 10 for other reasons. This then," he points to the chart, "is your Pareto chart."

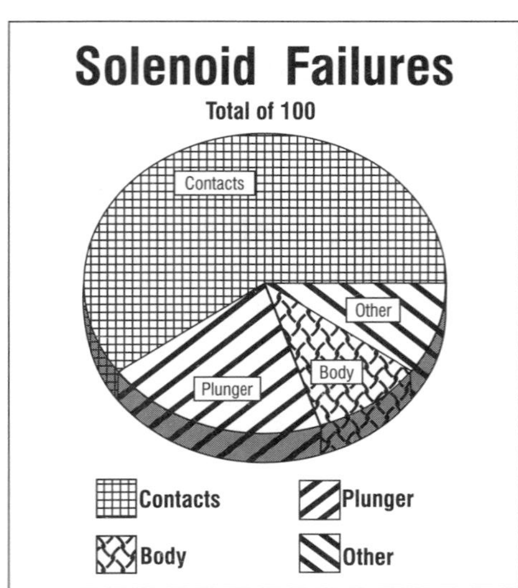

I have been writing like mad. It looks great. It seems pretty simple and I'm looking forward to trying it out. Dr. Stromberg is talking again. "You will want to look at your data a number of different ways. Number and type of defect, as I have done, is one way. You could also look at it using type of defect and cost to repair the defect. This gives you a Y-axis that is dollars and an X-axis of type of defect. It might be that one type of defect is 10 times more expensive to repair than others and this ought to be known before making any decisions. You might also choose time to failure or product model compared against type of defect. Looking at the information in different ways helps you learn more about your product, and the Pareto chart is a way of showing this information for easy understanding."

He draws another chart on the board. "I'm sure you have seen these kind of graphs before. A pie chart shows the same data, but you get a better idea of how big each segment is out of the whole. It is used widely because of the visual simplicity and power it portrays."

I have a couple of pages of notes. Most others have not written much down. I guess knowing you may need to use the information really increases the motivation to learn. As I'm finishing up, a girl a couple of rows in front of me asks Dr. Stromberg what Pareto means.

"Good question." He moves back to the board again and writes "80-20" beside the Pareto chart. Backing up a little, he asks the class, "Have any of you ever heard of the '80-20' rule?" I have and raise my hand and I see about three fourths of the class do the same. Evidently a number of people have heard about it. Dr. Stromberg continues, "Can someone explain it to me?" No one raises their hand at first.

Then a girl in the corner, the one who came in late the first day, speaks up. "The way I heard it is that 80 percent of the problems come from 20 percent of the kids in school."

Some laugh and Stromberg nods. "That is basically it. The concept was developed by the Italian economist Vilfredo Pareto describing the frequency distribution of any given characteristic of a population. He determined that a small percentage of any given group (20%) account for a high amount of a certain characteristic (80%). For example, eighty percent of the effects, whether it be defects, problems, sales, etc. come from twenty percent of the causes such as products, high school kids, or product offerings. This is very helpful information to know because if you can identify and remedy the 20 percent, in the case of defects or problems, then you can make significant progress in a desired effect."

Now Dr. Stromberg looks back at me. "Is that helpful?"

"Yes," I reply. "Thanks very much." I really am anxious to get to work and see how this will look on paper with my project.

He continues to cover the scientific method. Though I am anxious to try out my newly acquired knowledge, I try to stay focused and take more notes, knowing now that this stuff is more important than just getting a grade on a test or homework.

Dr. Stromberg continues. "A normal and frequent part of engineering work is the activity of experimentation and testing. You have been involved, as I have already described, with the plant and water example, in some single-variable testing and experimentation in your education already. You will learn other methods of experimentation that emphasize the inter-related nature of variables in any process or testing procedure. But that is in later classes and exercises." He puts a transparency on the overhead projector then stops and faces the class again as he continues to talk. It is a figure showing the procedure of the scientific method.

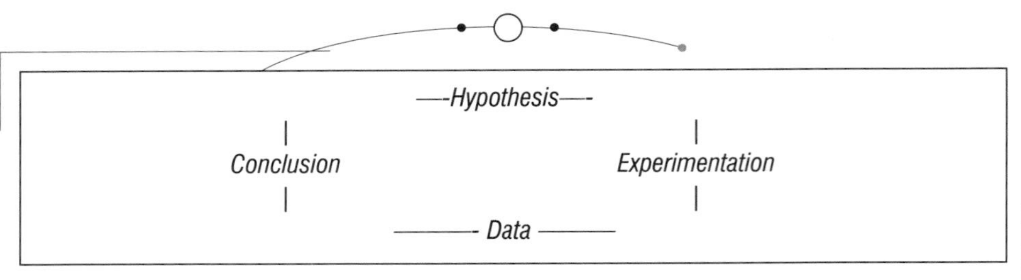

"Even though it is called the scientific method, the basic structure of the method is a very helpful process in gaining more information and understanding about an engineering principle or a problem. It is a constant cycle of investigation through experimentation, analysis, interpretation, conclusion, and recommendation which starts a new cycle."

He points to the sequence described on the overhead. "As you see here, the sequence is first hypothesis, then experimentation, then data. In reality, there are some cases when an experiment and the resultant data lead to a hypothesis that is then tested through further experimentation. This is still the scientific method and the desired result. That is, to learn the true nature of things is still realized."

Stromberg continues. " Let's talk about science and engineering. The difference between engineering and science is not in the use of the method, but in the purpose of the investigation, as described by Mr. Prados." He replaces the transparency with a new one with a quote from John Prados.

> *The goal of scientific activity is knowledge, an understanding of the physical universe in which we live. The goal of engineering is to create a device, system, or process that will satisfy a human need. The engineer may make extensive calculations based on scientific principles, but the purpose of these calculations is not to gain a better understanding of the principles; it is to obtain information on which to base a decision.*
>
> John Prados

Then he continues to speak. "Both basic scientists (physicist, chemist, etc.) and engineers gather and interpret data. However, the basic scientific goal is to discover and formulate a basic principle of nature. The engineer might use fundamental principles as a guide, but the basic engineering purpose is to understand the data in a way that will benefit mankind, and usually solve a practical problem."

Dr. Stromberg picks up a marker from the whiteboard tray and draws the cycle of steps on the board that were on the previous overhead transparency. He describes the steps as he does.

"Using the scientific method, you can begin with a hypothesis or begin with experimentation. When you begin with a basic assumption from which you will generate a hypothesis, then you determine some way to test the hypothesis and gather data. This is the experimentation step. A strictly followed procedure is critical while experimenting. Any departure from the procedure that you develop will introduce variation into the results. Those variations will lessen the accuracy of your conclusion, or even lead you to the wrong conclusion. The last step is to carefully analyze your data to determine what you have learned. From this, you draw recommendations. A simple, but sound procedure."

Dr. Stromberg stops writing and faces the class. "Now, sometimes you may just be curious about a certain effect or phenomena. In such cases you normally will not develop a hypothesis. Rather, you will start with experimentation and let the data lead you to a conclusion. Some have expressed concern that when a hypothesis is developed, the desire to prove it might lead the investigator to ignore data that doesn't fit. This is a danger, and it is related to the comment made by Doyle that I shared earlier. On the other hand, not having a hypothesis might cause the investigator to have no direction, and therefore, come to no conclusion. Still, it is helpful to just try things sometimes out of curiosity. How do we know when to do one or the other?"

Now Dr. Stromberg walks over to the table in the front of the room, and half sits—half leans on it in a posture we are starting to get used to. He says, "Let me tell you a story. This may help clear up the issue." This, I am learning, is his way of using fun stories and examples to illustrate his points.

"As you work in engineering and technology you are constantly learning. The experiments and tests you do add to that knowledge on a regular basis. This development process is not only part of the great enjoyment and satisfaction of engineering work, but helps keep the body of knowledge fresh. Hundreds of years ago, the development of what we call engineering was in large part accidental. Nowadays, we call these fortunate accidents serendipity, and they still occur. But for the point of the story it doesn't matter. A story of early sword makers is interesting with respect to the development of the technology of sword making. As early craftsmen made swords through the systematic process of heating, tempering and hammering, the story goes that one day a little redheaded slave boy urinated on a heated sword that had been laid aside to cool. It turned out that particular sword was much better than other swords. Other stories claim that heated swords thrust into the belly of a slave would be of much higher quality than swords that had not. Now, maybe the urine of redheaded boys is just the thing to make a sword great. I have my doubts that it is better than other treatments, but I don't know. The point is that a good engineer today, having stumbled on these important facts, would set up carefully controlled experiments to test these hypotheses instead of trying to buy up all the red-haired boys or lose a slave per sword. Remember, the real cause and effect is what you are after here, not proving your point that red-haired boys have a better effect than other boys."

He stops as if he is done. I see a hand go up and without waiting to be acknowledged, the student asks, "I'm still confused about when to use a hypothesis and when you don't."

Dr. Stromberg stands up straight again. "Yes, I didn't clarify that well, did I? In the case of the sword, you may state a hypothesis that swords dipped in the urine of redheaded boys are stronger than those that are not. You then develop an experiment testing your hypothesis with strength of the sword being the factor being measured. After conducting the experiment and gathering the data, you can then draw a conclusion proving or disproving the hypothesis. Now, let's say that the hypothesis is proved to be true…that is, swords dipped in urine are stronger. Sometime after, due to the difficulty of gathering urine of redheaded boys, you become curious if other urine, or maybe even other liquids, would give the same result. Therefore, you set up an experiment with a number of different liquids and run some tests. Your curiosity leads you to just test a few things and find out what happens. Your results might lead to new information or perhaps might lead you to develop another hypothesis, or might lead you nowhere.

Dr. Stromberg looks at the class and asks, "Is that helpful?" A number of people nod, and no one raises their hand.

He makes a few more closing comments about the scientific method, but I am thinking again about work, and how both the scientific method, and Pareto charts will help my work on the S-line products. I want to get out of here so I can get some lunch and head into work. I decide to skip my usual dose of ESPN afternoon sports and head to work early. Finally Dr. Stromburg stops, and I am gone.

I've been working almost an hour on finishing up reviewing the failure analysis sheets for the solenoids when Vic comes in. He looks at me, says hi and starts to walk out. Suddenly he turns around. "Hey, I thought you didn't come in till 2. Is it two o'clock already?" He looks at his watch, sees that it is not and looks at me again. "Can't wait to get to work huh?"

"Well," I stop my work and turn in my chair, "I am pretty anxious to get this done. I am really getting interested to see how it turns out."

He smiles. "Some projects really do get to you, don't they? Well, I am glad you're getting it done. When do you expect to finish reviewing the sheets?"

I am a little concerned about committing to a date but do anyway. "I think I can finish reviewing all the sheets today. I hope to have it all added up today, too."

He smiles again. "Well great. I am really interested in seeing the results. But I have to tell you, I am sure you'll find spring failure is our biggest culprit. Having the proof will be nice though. Maybe I can get some attention on those darn springs. I'll see you in a couple of hours." He turns and walks out.

As I listen to his opinion on the springs, I have to wonder if Vic will be right or not. I had started to form some of my own opinions but, after learning what I did this morning, I want to wait to form an opinion. Even so, it seems I haven't seen that many more spring failures than I have some of the other errors. Just then I hear Amy's voice talking to Vic as he leaves.

"I see Drew is here already. What's he up to?"

I hear Vic's voice fade as he hurries away, "He's up to showing that my theory on spring failure is right."

"I'm brand new here," I think. "I don't want to get fired already by showing a result different from what Vic anticipates." I decide I can't worry about it and just get back to work. I am as excited to try out the Pareto chart idea as I am finding the causes of failure.

"I don't believe it!" Vic doesn't look mad, just shocked. "This is impossible." He repeats it under his breath still looking at the Pareto chart I made. Yesterday when I finally completed recording all the data, I added up the counts and made a chart. It doesn't add up to what Vic thought it would.

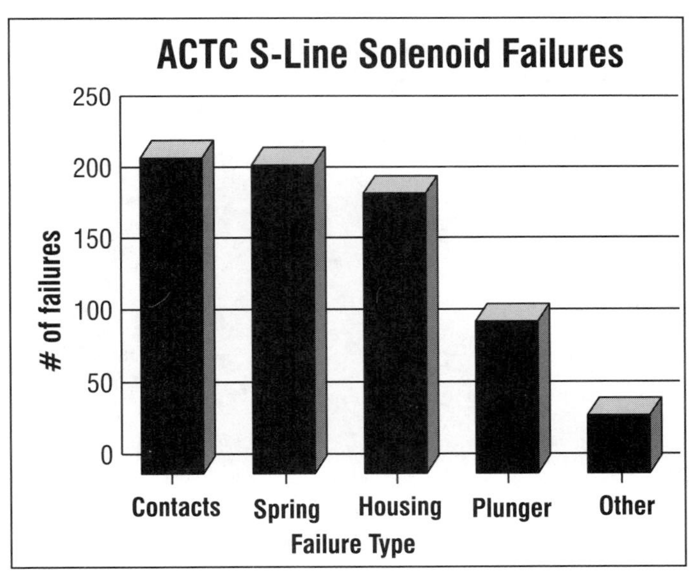

Amy is looking at the chart with Vic now. She looks at me. "Drew, are you sure this is accurate? Did you miscount these by any chance?" Amy is always calm and, at first impression, she almost seems uninterested. But that is just her nature. I nod my head as I answer.

"It's accurate. After I got the first counts I went back through and checked a number of them. I know it is not what we thought it would be, but it is accurate."

Vic has been looking at the chart since we have been talking. He finally speaks again. "I just don't understand how this can be so. I know that springs are the major cause of failure. I'm not doubting your work, Drew, but it really doesn't make any sense to me. In fact, just go out to the testing area and look on their breakdown bench. There are eight solenoids that came in last week, and seven of them have broken springs. Maybe that is a small and limited sample, but I see that a lot. Something just doesn't add up."

Amy speaks up again. "Drew, I don't doubt your work either, but why don't you take some time and check your counts again? Then maybe even go out to testing and show your results to Dan Johnson. He might have a thought on this."

Vic mumbles something about a meeting and walks out. Amy smiles and gives a word of encouragement and follows him. I sit down in my chair and look at the chart they left me with. I know I counted it up right. Now I get to do it all over again. Fortunately, I put the failure reports into piles according to the failure type as I went through them the first time. I'll just have to go through, check the failure type of each report and make sure I have it in the right pile. Then I can count up the piles.

I wonder if I ought to go see Dan in testing first, but I decide to go ahead and recheck all the numbers. I don't want to ask a dumb question if my numbers are off by a lot. I spend most of the rest of the day double checking the failure reports and recounting the piles. I finish as it is time to go home. My numbers are correct with two exceptions. Two failures counted as "other" were not, but instead were broken wire and burnt contacts. However, no change on the broken spring counts. I'll check with Dan Johnson tomorrow. I start to leave but the thought hits me to take the charts with me and maybe ask Dr. Stromberg his advice. I don't know if it will help, but what the heck. I put the charts in my pack and head out. "I hope this isn't against company policy," I think.

I'm still thinking about all this stuff as the bus comes. I know Vic has been here a long time, and if he feels that strongly, there must be something to his opinion. Maybe this is the frustrating, yet exciting, part of problem solving. I don't feel discouraged. In fact, I am excited about tracking this down. It is almost like being a detective. You take a lead and follow it and see what you get, then search for more leads. I hope Dan Johnson has a new lead for me tomorrow.

Chapter 9

For some reason, I can't get the S-line spring problem out of my mind. Last night I went out for ice cream with Christy and found myself telling her about it. I was really embarrassed, but she was great about it. She joked about guys using all kinds of lines on her before, but never an S-line. I'm anxious again to get to work and talk to the guy in solenoid repair, Dan Johnson. In the meantime, I had better get to class.

When I get to class Dr. Stromberg has an old sword sitting on the front table. I hear someone behind me joke that Stromberg is looking for a redhead to help him finish it. I glance around and see that the guy the comment is directed to is a redhead. He is flushed but grinning. As soon as Dr. Stromberg begins talking the class quiets quickly. I guess we are learning. "Please don't worry," he says, smiling now. "This sword is finished, and I don't need any redheads to do anything to it." Everybody laughs and the redheaded kid behind me gives a fake sigh of relief. Everybody laughs more at him. He seems pretty good natured.

"I brought this sword today as an example of some of the ingenious inventions and products of the past. Too often, we think of ourselves as the most advanced people ever to walk the earth. With respect to some things, such as computer technology, this is correct. However, we would be gravely mistaken if we believe we are smarter or more advanced in terms of thinking than any other society. There have been many societies that have performed some great feats of discovery and advancement. I want to share some of those with you."

He turns on the overhead projector, and pulls a transparency from his folder but doesn't put it on the projector. He pauses for a moment, then continues, still with the transparency in his hand.

"What is the work of an engineer like? Why would one want to do engineering, and what motivation is there for the work of engineering? Anybody here heard of a man by the name of Herbert Hoover?" He stops and looks around.

Nothing happens for a minute, then someone on the right side of the room raises her hand. Dr Stromberg nods at her, and she says "He was president of the United States sometime, I think."

"That's right. What was his career before politics? Anyone know?" Nobody volunteers this time, so Dr. Stromberg continues. "He was an engineer. Trained in mining as I recall, but educated and worked as an engineer." He puts the transparency he had first taken out of his packet back, and then picks up a separate folder. Looking through it for a moment he takes a transparency from it.

"This might be of interest to you. This is a statement by Herbert Hoover. It is a little long, but let's read it." He puts the statement on the projector and reads it for us.

"The great liability of the engineer compared to . . . other professions is that his works are out in the open where all can see them. His acts, step by step, are in hard substance. He cannot bury his mistakes in the grave like the doctors. He cannot argue them into thin air or blame the judge like the lawyers. He cannot, like the architect, cover his failures with trees and vines. He cannot, like the politicians, screen his shortcoming by blaming his opponents and hope that the people will forget. The engineer simply cannot deny that he did it. If his works do not work, he is damned. That is the phantasmagoria that haunts his nights and dogs his days. He comes from the job at the end of the day resolved to calculate it again. He wakes in the morning. All day he shivers at the thought of the bugs which will inevitably appear to jolt its smooth consummation. On the other hand unlike the doctor, his life is not among the weak. Unlike the soldier, destruction is not his purpose. Unlike the lawyer, quarrels are not his daily bread. To the engineer falls the job of clothing the bare bones of science with life, comfort, and hope."

—Herbert Hoover

Dr. Stromberg looks back up to the class. He is smiling now. "This is the glory and the damnation of an engineer. To invent, design, produce and improve on a daily basis for the benefit of society and the advancement of technology is exhilarating. Certainly, as President Hoover describes, it can be a frustrating and daunting task. It is also, however, an invigorating and exciting opportunity. To see something succeed that has been very nearly the sole occupation of one's mind and work is euphoric indeed." He starts to take the transparency off of the projector, but a voice from behind me stops him.

"Wait Dr. Stromberg. Can you tell me what," speaking very slowly now the student tries to say, "phan-tas-ma-gor-ia, means?"

"Good question. I'm sorry, I should have explained. In this case, phantasmagoria means something like an optical effect that makes an image appear to diminish into the distance or come toward the observer with increasing speed. Perhaps you have seen a video segment where something starts far in the distance and with increasing speed comes closer until it appears to run over you. That is a phantasmagoria. Here President Hoover is telling us that the engineers' work is like this diminishing then enlarging optical view of things. Like a dream, or perhaps a nightmare, that constantly occupies your mind."

He looks up at the student and asks, "Is that helpful?"

The young man nods but then adds, "Yeah, but it seems kind of scary. I don't know that I want to have nightmares about my job."

Dr. Stromberg chuckles and answers him. "Well, anyone who is deeply involved in their work will have this kind of experience, engineering or not. President Hoover is, as I said a moment ago, describing the glory and damnation of engineering work."

The thought suddenly hits me like a rock. Is this what has been going on with me the last few days? The S-line failure problem has seemed like it has been on my mind almost constantly. I even had a dream last night about springs in solenoids breaking. It makes at least some sense to me what he is saying. I am sure many people experience much greater feelings about this than I am right now. The interesting thing is that it isn't a negative thing. In fact, it is kind of exhilarating, like Dr. Stromberg said.

Dr. Stromberg is still talking. "Engineers and technologists have a significant impact on change in the world. A large part of improvement in standard of living and progression in society happens in engineering and technology." He stops to see if there are any questions or reactions, but no one responds.

"Well," Dr. Stromberg walks back to the front table to get an overhead transparency from his packet. "I wanted to talk a little about the benefits of a career in engineering and technology today. Our discussion to this point certainly fits into that. Let me show you some other information about this area as a career and lifelong practice." He takes a transparency out and places it on the overhead projector. It is blank except for a heading, "Benefits of a Career in Engineering and Technology." I hope this doesn't mean there are no benefits. Dr. Stromberg takes a marker from his pocket and prepares to write on the transparency. He looks to the class and asks, "What benefits do you believe there are to a career in engineering or technology?" He waits a few seconds and there are no answers, so he asks again. "Why are you interested in this field? Just call out some answers."

Someone finally calls out, "The engineering people I have met seem to like their jobs."

"Great," he says and writes "Job satisfaction" on the transparency.

That breaks the ice a little bit and another student chips in. "It seems like there are pretty good jobs in this field."

"Another good one," Dr. Stromberg says as he writes "Employment security" on the transparency. A few more students call out some things and soon he has a list of a half-a-dozen benefits written on the transparency. I'm glad he wrote "good pay."

Dr. Stromberg looks at the list then adds two more so now there is a list of eight.

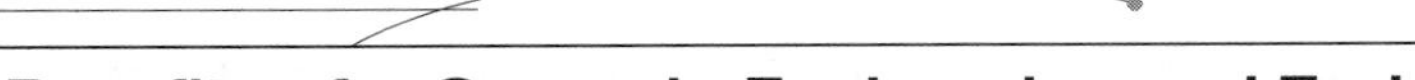

Benefits of a Career in Engineering and Technology

Job satisfaction

Employment security

Good pay

Challenging work

Status

Being able to understand how things work

Contributions to society

Individual and career growth opportunities

As Dr. Stromberg steps back and looks at the list, I look at it too. This is pretty good stuff. Not that other careers don't have some of these same things, but seeing some of the benefits listed like this is impressive to me.

Dr. Stromberg moves toward the middle of the room and begins to talk. "This is a great list. Let me expound on some of the benefits you have listed here because I'm not sure we appreciate as well as we should how important some of these things are. For example," he points to the first one with a laser pointer he removed from his pocket. The pointer is kind of neat. It looks like a pen and instead of a red dot it projects an arrow. He continues, ". . . more people are dissatisfied with their work due to lack of challenge and growth than any other complaint. Work in the field of engineering offers endless challenges, and the environments in which engineers work are normally quite comfortable and highly professional. There is no end to the improvements that can be made in products, processes, and systems regardless of the field. Things that were thought impossible just a few years ago are standard today. The same will continue in the future as advances in technology open new vistas of opportunity for virtually every field of engineering."

He draws a circle around the first four things listed and then continues. "Though people in many different careers will have the chance to use advances in technology, those involved in engineering will always be on the cutting edge and be part of the development and implementation of state-of-the-art technological solutions. These opportunities will provide high job satisfaction and a challenging work environment. In addition, the salaries that engineers receive are very good, and job opportunities are good as well, even in some slow markets."

He puts another overhead on the screen. This one is a table showing some salaries. "Engineering graduates have some of the highest starting salaries of any four-year profession. Table one shows the yearly starting salaries for various professions. Engineers also tend to enjoy a high degree of employment security. They tend to be able to hold jobs well and have good opportunities open to them, if they do desire to change."

I realize that Dr. Stromberg is doing kind of a sell job here, but this is sounding better than I had thought. I'm impressed with the starting salaries though, because I didn't know they were that high. I also hadn't thought about some of the other things he is telling us. Then I look at the rest of the list on the transparency and it hits me that I have thought even less about some of those things. I wonder what he will have to say about them. He starts to talk again but then sees a hand up in the room. He asks for the question.

"Dr. Stromberg, are you talking about engineering and technology in general now? What area within engineering and technology would be the best? I know where I think I want to be, but maybe there are better jobs or more pay in another area."

Dr. Stromberg takes off his glasses and ponders for a moment, then continues. "That is a question that often comes up and I really believe the answer is always the same. That is, all of the areas of engineering and technology are wonderful opportunities. Specific disciplines and fields within the general area of engineering will vary slightly both in terms of salary and number of jobs, but in the long run, you will be happiest and best served to pursue the area you are most interested in and which best fits your talents and gifts."

While it sounds like an answer I would expect him to give, it also makes some sense. Dr. Stromberg is speaking again. "For example, we did a survey of our alumni a year or two ago asking various questions. One of which asked how pleased the graduates were with their degree. The large majority were very pleased with their jobs and the degree they chose. A few indicated they wished they had chosen a different field in engineering or technology than they did. Perhaps they had chosen primarily based on where they thought the money was. Most, however, had found their way to where they wanted to be. One I felt bad for was a man who wrote that he had disliked his job of 15 years. He said he had gotten into his specific field because of good job opportunities and because all of his brothers had gone into it. He wrote, "The job opportunities and the money are great, but I don't enjoy my work. I wish I would have pursued the field I always wanted.' What he had always wanted was not even in engineering and technology." He looks up at the student and says, "That doesn't answer your questions specifically, but I hope it helps."

Dr. Stromberg walks to the side of the room, points his laser beam at the next four items listed, and begins talking again. "The study of engineering provides an excellent way for you to learn to think well, develop intellectually, and contribute to society. Engineers are trained to look at problems and tasks in a logical and organized way. The abilities and skills you learn in your engineering education and practice will be valuable in all areas of your life. You will be better equipped to contribute to your community, school, church, and other organizations through the knowledge and expertise you gain. Your knowledge of basic natural and physical laws will also provide an insight into non-technical problems. For example, knowing cause and effect in relationships, you will be more apt to see through smokescreens of rhetoric and ill-founded ideas or solutions."

I had never considered this side of things. I really wonder if this is true. Still being a teenager, I haven't considered what impact I can have other than getting a job and making some money. These ideas are kind of interesting and the way he is explaining them make sense to me. I hear Dr. Stromberg is still talking and start listening again.

"Unfortunately, engineers have not been as significantly involved in community, state, and national issues as they should be. Society needs not only technical solutions but also solutions to many social problems. Even though the technical skills you obtain may not be directly applicable to the social or environmental ills of the day, your ability to analyze problems, generate ideas, and implement and evaluate solutions is greatly needed and wanted. We hope you will see your talents and the skills you learn as beneficial to more than just the company you work for. Your intellect will sharpen as you work on problems outside your specific area, and your satisfaction will increase as you contribute to many areas of society outside of work."

Now he points at the item that mentions being able to understand how things work. "The training you will receive, combined with your inherent talents and abilities, will help you understand how things are built and how they work. It is very satisfying to look at buildings, roads, automobiles, household items, airplanes, and thousands of other items and be able to understand how they were built and the principles which on they operate. Additionally, if you are interested, the knowledge and skills you gain in engineering will allow you to fix almost anything, bringing satisfaction and confidence, not to mention saving some money. For example, I very seldom call a repairman to my home. I am able to fix things in and around my home, and enjoy the time spent doing it."

He now shuts off the overhead projector, again takes off his glasses and puts the laser pointer back in his pocket. I glance at the clock and realize the time has passed faster than I thought. There are only a couple of minutes left in class. This, I think, is how I like class to go.

Noises of pack zippers and notebooks closing begin to echo throughout the room. At the same moment, we all realize that Dr. Stromberg has not excused class yet and is just looking at us. Class quiets down again. He nods his approval then says, "These are just a few of the benefits of being involved in engineering and technology. There are many more. It is a very respectable, exciting, and valuable profession. Now for this week's assignment write a two page paper. . . ." he stops and smiles for a moment as groans come from all corners of the room, then repeats, ". . . write a two page paper on 'When I graduate, I want to be an engineer/engineering technologist because . . . .' Then you fill in the blank and finish the sentence." Almost immediately a hand goes up and I bet I know what the questions is. It looks like Dr. Stromberg does too. He holds his hand up as if to ask us to stop, and says, "Now some of you might say, 'I don't know what kind I want to be, or why, or even if I want to be.' I understand that. Pick one anyway and write the paper. The purpose is to have you think about why you would." He lowers his hand, looks at the student who raised his hand and asks, "Do you have a question?"

"Not anymore," he replies. "Thanks." Dr. Stromberg excuses the class and pandemonium breaks out like every time a class is over. This assignment looks kind of fun, even though I am not sure what kind of engineering or technology I want to go into. I think I can come up with some pretty good ideas. I'll ask some of the others at work too. I'm sure glad I have my job. It's coming in real handy. As soon as I think about the job though I start thinking about the S-line project again. I can't wait to get to work and talk to Dan Johnson. Right now though I am going to get some lunch.

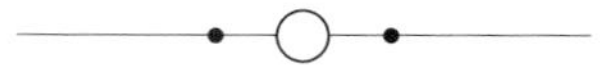

As I am eating lunch, I remember I have the charts from work and want to see what Dr. Stromberg thinks of them. I hurry and finish and go to his office. The secretary says I just missed him. I guess he is going to be gone for a day or two.

I get mad at myself for not having remembered right after class. I guess I'll have to wait. I have no choice really.

I go back to my dorm and get hit with a surprise. I check my mailbox on my way to my room. There is a letter from housing announcing a raise in rates effective beginning the first of next term. I had just figured out I would have enough money to take care of tuition, books and housing. Of course, I want a little left over for fun and such. Now this increase really puts a cramp in my plans. I hope the money comes up from somewhere. I decide not to sweat it now. Besides, I still have almost a month and a half before the money is due. It kind of makes me mad though. Seems like the price of everything is going up. I wonder how Mom and Dad do it, especially with Dad's illness.

When I get back to my room, Stu is there. It looks like he is actually studying. He looks up when I come in. "Hey Drew, how you doing? Listen, I just talked to Mom. She said hi and Katie said hi too. Hey guy, I think she likes you." He gets this real serious look and says in a deep voice. "Young man, I want to know what your intentions are with my sister." Then he laughs and changes the subject. The good thing about Stu is that if you don't want to answer his question or don't like what he is saying, just give him a minute, and he'll change the subject.

"What you doing now? he asks. "Studying?"

"Naw, I think I'll head out to work." I put my books down and grab some coins for the bus fare out of my coin jar.

"Work?" Stu sounds surprised. "I thought you didn't go to work till 2:00 or so."

"I don't usually, but there's a guy I need to see to help me on a problem I am working on. I want to make sure I get to see him today." I don't say anymore. I don't want to sound obsessed with this S-line problem.

"Still the S-line stuff?" he asks. I nod, I am surprised he knows. He must have been listening the other night when I was talking about the defect problem. Then, true to form, he changes the subject again.

"Hey, you're a real brain and an engineer. Let me ask you a question."

I want to tell him that first off, I'm not a brain, but I don't argue. "What is it?" I ask him.

"This physics stuff is driving me crazy. I've got to have this homework turned in this afternoon and I don't get it. Besides, why do I need physics? I'm in business, I'm not going to invent a new element."

I resist the urge to explain that new elements are in chemistry, not physics, and just sit down by his desk. He sees that I'm ready to help and continues. "This is the deal. There are two dots on this round plate. One is here," he points to a figure on his assignment sheet with a dot about an inch away from the center. "And one is here." Pointing to the other dot that is near the edge about 3 inches or so from the center. "I am supposed to tell the guy how fast each dot is going in inches per minute if the plate is spinning 25 times a minute."

"What class is this?" I ask picking up his workbook.

He grabs the book from me and snaps, "Don't worry about what class it is, OK? Maybe it's physics for dummies. I don't like science stuff, but I have to take something to get these GE's out of the way. Somebody said this was the easiest."

I sit up and hold my hands as if to surrender. "OK, OK, no problem. Let's just look at the problem. What don't you see?" I am still not sure what he doesn't understand.

He gets a frustrated look on his face. "Look, each one is going around 25 times a minute. It looks to me like they are both going the same speed, 25 times a minute. What more do you need to know?"

"Stu," I sit closer and point to the circle and dots, "speed, or better yet, velocity, is a distance over time issue, not how many times it goes around. Let me ask you this. For each time the plate turns, which dot has to go further in distance? This one or the one on the outside?"

He still look confused. "The plate is not going anywhere. It is just spinning. I don't see how any dot is going any distance."

An idea hits me. "Just a second," I tell him. I pull a dime from my bus fare out of my pocket. "Now take your pencil and trace a line around this dime and look at it. How long is that line?"

He draws the line and looks at it and says, "About an inch, I guess."

"Good," I tell him. Then I grab one of his pop cans from his shelf and place it on the paper. "Now do the same with this can and tell me how long the line is."

He traces the can, picks it up from the paper and says, "Looks like about six or seven inches." Then suddenly his face brightens.

"Oh, I get it. The dot out here has to go further for each time the plate goes around. You said speed is a distance thing, so it has to go faster than the other dot to stay on the plate. If it went slower the dot would fall off." I laugh out loud but it doesn't bother Stu. He has the "Eureka" look. He continues, excited by his new discovery. "So I just need to measure how many inches the dot goes in one turn, multiply it by 25 and I've got it." He looks jubilant.

Then the jubilance turns to puzzlement. He looks at me again. "I know there is a way to measure the length of the outside of a circle. What is it?"

"πd," I tell him.

"Wait, what?" He grabs his pencil. "Pi is that constant thing right?" I nod. "3 point something isn't it?"

"Good memory Stu. Yeah, it's 3.1416."

He writes that down. Now he is muttering to himself. "I know d is diameter. So the diameter of this one," he points to the dot closest to the center, "is 2 cuz it is 1 inch from the center. So that's 6.2832, then I times that by 25 which is 157.08." He looks at me smiling. "So the answer is 157 inches per minute, right?"

"Actually 157.08 inches per minute."

He laughs and mumbles something about engineers. "And the other dot is . . . ." he hits some buttons on his little TI-30 calculator, " . . 314" then he emphasizes "POINT ONE SIX. HA!" he yells as he finishes writing down the answers, stuffs the paper in his book and slams it shut. "You're great Drew, world's best engineer, that's what I tell everybody." Then he slips on his sandals and heads out the door.

"Where you going?" I ask him. He looks back and says.

"All this talk about pi got me hungry. I'm going to the cafeteria." I can't help but laugh as he runs out the door.

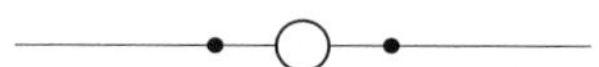

On the way to work I start thinking about Stu's frustration with physics class. To him, it is a general ed class to get out the way. To me, physics is an important part of what I want to study. To him, general economics is a key part of his study. To me, it is a general ed class I want to get out of the way. It dawns on me that general ed is, at least in part, to have each of us study outside our disciplines or areas of professional study in order to be more familiar with the world and subjects in it. Even though it makes sense, there are still some classes I would rather not take. Music theory is not my idea of fun. I also remember that Stan Hall told me to stop in some day and talk to him about "getting the GE's out of the way." Maybe I'll do that today. But first I've got to find Dan Johnson.

Chapter 10

"Yeah, I know what the problem is." His matter-of-fact tone surprises me. "The spring is the problem . . ." Oh no, I think, here we go again. But he continues talking ". . . .but only in the smaller S-line models. In the larger ones, the 300 and 400 models, it's not too bad. But in the small models, the 100 and 200 series, the spring fails more than anything else."

As he is talking, I realize that I have all the S-line models combined together in my failure analysis chart. I didn't separate them at all. I didn't think it would matter. Amy and Vic didn't say anything either when they saw the data yesterday. I wonder why. I finally respond to Dan.

"Are you sure? I mean, I don't want to doubt you, but how do you know?"

He stops what he is doing and points to some solenoids sitting on an adjacent workbench. "See those?" I nod. "Four of the eight have broken springs. Guess which four."

I look at the solenoids he is talking about. Seven of them are the smaller model, the 100 and 200 models. One of them is the 400 series. I venture a guess that will please him. "Four of these?" I point to the 100's and 200's.

"Yep," he says. "Those are four broken springs, two burnt contacts, and one cracked housing. The 400 is a burnt contact."

"This is just 8 solenoids. Are you sure it's true for others?"

He looks frustrated. "You have the failure reports don't you?" I tell him I do. "Then count them up yourself. I tried to tell some people a few months ago that the springs on the smaller S-lines had something wrong but nobody seemed to care." Then he looks at me suspiciously and asks, "Are you a college boy?"

I'm not sure what he means but I answer anyway. "Well, I am going to State University if that's what you mean."

"Then maybe they'll listen to you. I only graduated high school and did some time at the Vo-Tech. I'm supposed to test these things and that's it. Seems like they don't want to hear if I have an idea about how to make them better or what might be wrong. Anyway, you go count them up. If I'm wrong, I'll buy you a drink. If I'm not, then you just tell me. I'll be satisfied to have someone listen."

I thank him and leave, but just as I get out of the lab another question hits me. I go back in. "Dan, any idea why the small models have such a problem with the springs?" I'm hopeful that if Dan is right I can come up with a solution.

"No idea at all," he says, "looks to me like the same material. We make the springs in-house so there is no vendor differences. I haven't got a clue on that one."

"OK. Hey, thanks, Dan. You and Vic have just about the same idea, so there must be something to it. I'll let you know."

As I turn to leave, Dan says, "You do that. I want to know what the problem is too."

I hurry back through the shop to my desk. This idea of classifying them by models is a pretty simple idea. I should have thought about it. I remember Dr. Stromberg told us to look for logical divisions when we show data on a Pareto chart. This is a logical division. What doesn't make sense is that if the springs are made out of the same material why some would break and others not.

I've got to slow down here though. I don't even know if what Dan said is right yet, about that the smaller models having more broken springs. I had better check that out first.

When I get to my desk Vic is in his office. Fortunately, he is working pretty hard on something and only says hi when I pass him to go to my desk. I really don't want to say anything yet. I want to get these new charts done first. It'll take a while, because I need to go back through and separate all the failure reports by model number and by defect type. I decide not to do the 100's and 200's together but to make four separate counts. If the 100's and 200's are pretty close in number of similar defects, then it will be easy to combine them later.

I glance at the clock. It is just past 3:30. I can't believe the time has gone that fast. I hope I can get the counts done before 5:00. I can't stay late today. I have a lot of homework and tomorrow being Friday means a test in physics. I start right in on the reports looking forward to what this will show me.

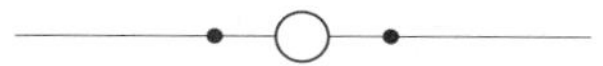

I didn't get the new counts all done. In fact, I only got about half done. Separating the failure reports by model took longer than I thought, then Vic came in and wanted me to attend a meeting with him. It was on the new S-line products. He thought it might be good for me to know what the plans were for the product. Turns out it might be a good thing I was there. They are planning improvements in all of the model lines, so if this little search I am on turns up something, then maybe it will be more helpful than we thought, particularly if Dan Johnson is right. Vic and Amy will both be gone tomorrow, so I'll be able to get the counts finished and chart them. I can't wait to get it done.

I'm still thinking about it at dinner. The same thing is on my mind and something hits me. What if Dan is right and the springs on the smaller models break more than the 300's and 400's? What then? I decided I can't worry about that right now. I have to take it one step at a time. Right now, the step I need to take is finish eating and then study for my physics exam.

Just then my thoughts are interrupted. "Hey Drew, how ya doing, guy?" I look up, and it's Troy.

"Doing well. Just finishing up dinner here. What is this stuff anyway?" I hold up a spoonful of what was dished on my plate.

"I don't know." Troy answers. "But I heard it was left over from World War II, and they wanted to get rid of it before it spoiled."

The guy at the next table about loses it as he bursts out laughing. "Come on, guys," he complains, "I'm trying to eat here."

Troy turns back to me. "How about we hit a movie tonight? There is a great one at the student center."

"Ahh man, that'd be great but you know that physics test tomorrow. I really need to study for it. I better not go," I tell him, knowing that I would rather be watching the movie.

"Oh come on Drew," he's coaxing me now. "It's an early movie and it will be out by 9. You can still be back by 9:30 and get 2 or 3 hours of study in. Besides, I really want to see it, and I don't want to go alone. Come on, you'll do OK on your test."

I waiver for a minute, but even in my waivering I know myself well enough to know that I'm giving in. Besides, he's right, I can still get in some good study afterwards and be ready for the test. "OK, let's go." I get up to take my tray back.

"All right," he's grinning now. "But aren't you going to finish eating?"

"No way. This stuff is gross. I'll finish my meal with some heavily buttered popcorn at the movie. It may not be as healthy, but it'll taste a heck of a lot better."

Off we head to the movie. I'm looking forward to it and I'm pretty successful suppressing the feeling of guilt about the study for the physics test. I figure I deserve this break. I've been working hard and studying pretty well too, so why not enjoy some time from those things? Even with the statement — "Rewards for completing a goal or task must support the goal, not be contrary to it," — from the class on goal setting I attended a couple of weeks ago echoing in my mind, we head up the hill to the student center. I give it very little thought after that. Entering the student center I see the movie is an action movie I've wanted to see too. This, I am sure, will be a good break. Besides, Troy was in the same goal-setting class and physics class. He wouldn't lead me astray.

The next day when I went to take my physics exam, I still felt pretty tired. I don't think I did too well on the test. I was a lot later last night than I expected to be. The movie was out by 9:00, but as we were leaving we ran into some common friends, and after a pizza and a lot of talk, it was well past midnight before I got back to my room. Even Stu was home before I was. I did cram for the test for a while but I was already feeling beat and couldn't concentrate like I should have.

Here comes the test score. 69%. Not what I had hoped by any means, but probably what I should expect. My only hope now is that I can have a chance to make it up somehow. I will need to adjust my study schedule. Russell is very good to help out. I'll take him up on his offer to join his study group. They meet every Monday and Friday at 6:00 p.m., which is certainly a time I can arrange. I'll make the adjustment on my study schedule when I get home.

I head back across campus to get to my English class. It has been going pretty well but writing has not been that hard of a subject for me. This particular class is a freshman writing class required both by general education and engineering. I need to come up with something to write about that is related to my major and would be interesting to someone outside my major. I already figure I will do something that will be related to what I am doing at work. I figure something on this project with the solenoids will work for the assignment.

In English class I can't concentrate very well. I'm anxious to get out of the class, get some lunch and head to work. I really want to finish the data on the failures and see if Dan Johnson is right about the springs of the smaller S-line models.

As I am walking in to work, I see Stan Hall. He still remembers me from that first day. I hope this is a good thing. "Hello Drew," he says extending his hand. I take his hand and shake it answering him. He continues to talk. "When are you going to stop by and see me?" He asks. "You do remember that I asked you to, don't you?"

"Yes sir," I reply, "I just know you are busy and I don't want to interrupt you." I am half lying. I really kind of thought he was not serious but just being nice.

"Well good. I want to see how you are and just share some ideas with you about how to become a good engineer. I hope you will stop by. As a matter of fact, why don't you do it next week, maybe Tuesday or Wednesday. I'll be in town all week and I would look forward to visiting with you. Will you do it?"

He asks like I might say no. I may be dumb, but I am not *totally* stupid. "You bet I will, Mr. Hall. I will be there next week. Any time better than another for you?"

"I don't know. Just stop by and check with the secretary on one of those days and see what she says." By this time we are in the building. "See you next week." He pats me on the back as he heads to his office. He takes a left while I continue down the straight hall to my desk.

"Yes sir, I will be there." I actually think it is really neat to have a chance to visit with someone who has made a name for himself like this and is successful in the company. It is also kind of scary, though. I'm trying to remember what it was that made him invite me to visit with him. I think it was something about getting through the general ed stuff in the programs. Yeah, that was it. I was thinking he might have some advice on how to get it out of the way. Guess I'll find out next week.

By this time I'm to my office. Well, not really an office I guess, just a desk. Vic's gone, and I remember that he and Amy are both gone today. Something about visiting one of the suppliers for the S-line product. He has left me a note. I pick it up.

> Drew:
> Amy and I have gone to MetalPro to talk to them about some new material they want to use for the housing on the new S-line. I know you are still working on the failure reports. If you get done and need something more, go ask Dan Johnson if he has finished the failure tests on the last returned solenoids. After that, if you have some time, get the CAD package training manual on my desk and start reading it. We want you to learn the package so we can have you help us generate supplier-ready drawings for the new products. Thanks, see you on Monday.
>
> Vic.

Well, getting the other failure reports from Dan is something I need anyway. Besides, I already have a pretty good idea of what they will say. He told me as much when I was there yesterday. The CAD manual stuff sounds like it might be a little bit of busy work. I thought you only got busy work from the professors at school. I didn't know it would be here too. I catch myself though. I better be careful. Just because I don't want to sit and read a user manual for a computer doesn't mean it is not important. I decide to worry about that later. I've got to get these charts and figures done.

Two hours later, I decide I will go see Dan Johnson. He has some reports I need, and I think he will be very interested in what I have found. I think it's amazing how this looks.

"Hey Dan." I knock on the door jam as I call him.

He turns. "Yeah, what do you want?" He sees me. "Oh, hey college boy, it's you. How's your work on the failures coming?"

"That's what I want to show you." I tell him. "I was confused by the first chart I did. You are not going to believe this." I show him the new Pareto charts I have generated. Then I look at him and say. "So what do you think?"

He looks at me and says. "Well I'm not a college boy so I don't know what all your fancy charts mean . . .." Then he looks back at the reports and continues, "but it looks to me like these" he points to the charts of the S-line 100 and 200 charts, " . . . are mostly springs breaking." Then he points to the other two charts which are the 300 and 400 S-line model failure and says, "And these are just a lot of different reasons but nothing in particular."

"You hit it just right, Dan. Isn't it great? You were right." The look he gives me tells me that he is not as surprised as I am.

"I told you that yesterday," he says matter-of-factly. "What's wrong, didn't you believe me?" I realize I better be careful so I try to back up a little bit and explain what I have done.

"I didn't mean that I didn't believe you. I wasn't sure what to believe. I just needed to check all these numbers. When I was here yesterday and talked to you I realized that I had combined all the numbers together. Here, look at this chart I did before." I show him my first chart I did. "The type of failure is down here." I point to the X-axis. "You can see the springs, cracked housings, burnt contacts etc."

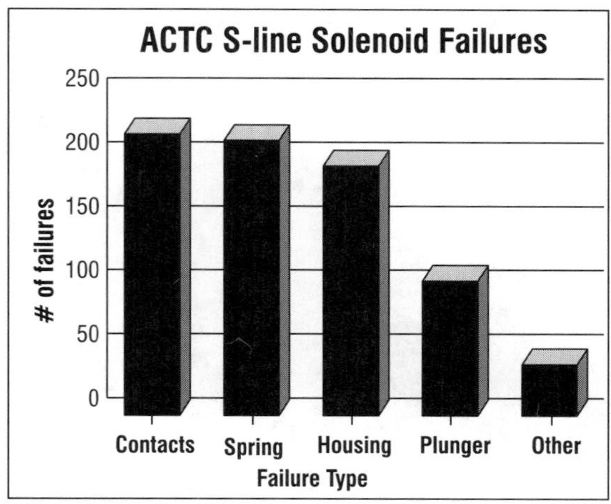

"And the number of failure of each type is here." I point to the Y axis. "All of these add up to the total of the 736 total failures.

I continue explaining. "This looks like no one failure is happening more than any other."

He nods, "Yeah, overall all that is probably right."

"OK then," I tell him. "Then yesterday when you said that the 100's and 200's had a lot more spring failures than the 300's and 400's, I realized I had grouped them all together and that I shouldn't have. So, I went back and counted failures per individual product type, and this is what I got for the smaller models.

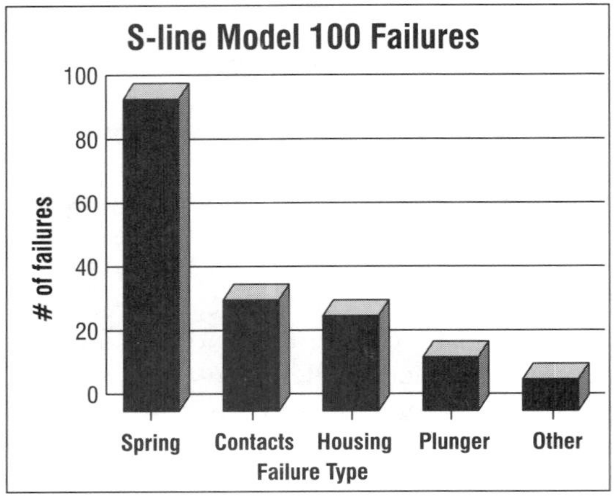

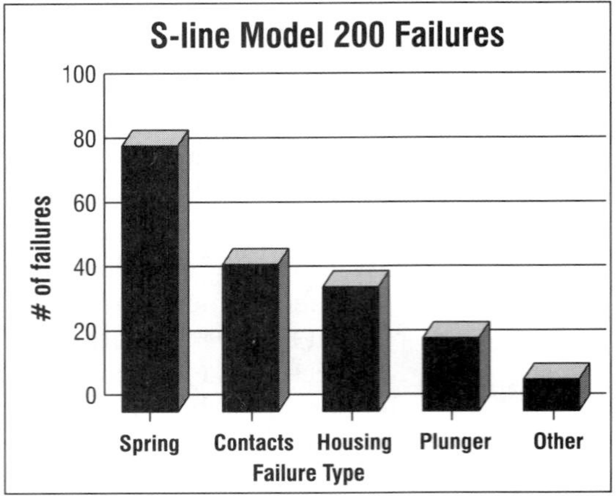

"Then the larger models look like this."

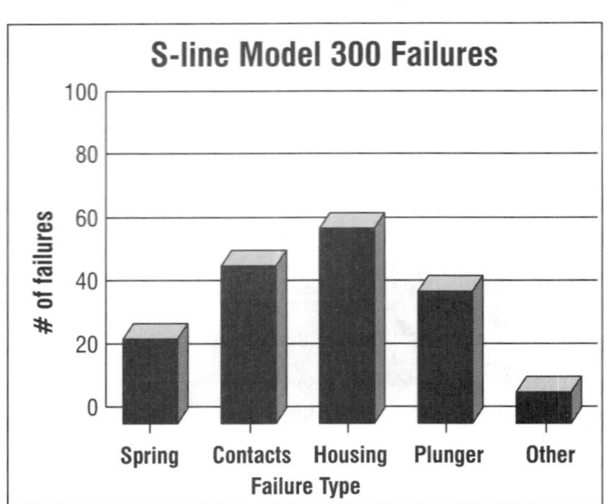

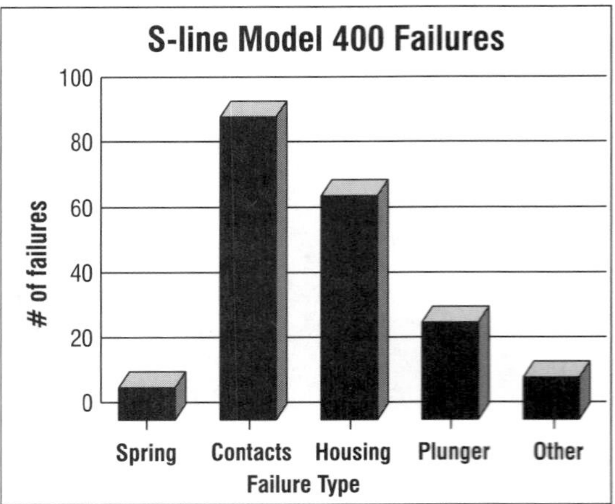

He looks at them for awhile then looks at me and asks, "You mean you did these reports because what I said made you think to look at them that way?"

I'm a little surprised by his question. I thought he would be pleased to see the results. I answer anyway, "Well, yeah, I guess. I mean I wouldn't have thought to break them out separately if you hadn't said what you did. Is something wrong with that?"

He looks a little embarrassed. "No, course not. I didn't think you college boys would listen to me. That's all."

"Listen Dan, first of all I just started school here; I'm just learning some stuff and don't really know what I am doing. Second, you do this all the time, and Vic is the one who told me to talk to you because you would know what you are doing."

He now looks surprised. "Really?"

"You bet. Why?"

"Oh I just didn't think anybody figured I knew too much, that's all. Oh, hey, listen, I don't mean no harm calling you a college boy. I think it's great you are going to school. Sometimes I wish I had done more school. It seems kinda late now though, with a growing family and all. But others are doing it, maybe I can too." Suddenly he stops talking and straightens up. He then asks quickly. "So what does this mean? Do you know what is wrong with the small models?"

"I wish I did. All I have done is prove that what you believe is correct, and for that matter what Vic believes is right. I have no idea why they are this way. In fact, I was hoping you would have some idea. Are the springs for the small models and the large models different in any way except size?"

"I don't think so. In fact we have our own machines. All the springs are made here." There goes another idea I had.

"Are you sure? Are any of the sizes made by outside people?"

"Nope, I am sure of that. Jess over in springs is a good buddy of mine. We hunt together. He runs some of the spring machines, and I know we make them all here. We have ever since they started using the new Springalloy material. They decided that was a good time to bring the spring production in house."

"Any ideas at all about this? There has to be some reason why these are so different."

He shrugs his shoulders. His eyes light up a little. "Hey, Jess might know though. He has been there awhile and is pretty good at it."

"Great!" I exclaim. "Can we go talk to him?"

Dan shakes his head. "Not today. He glances up at the clock. "He leaves at 3:30." I look at the clock and am surprised to see it is a couple of minutes to four.

"How about tomorrow?" I ask.

"No can do then either. We're taking tomorrow off and going duck hunting. I'll tell him you want to see him and maybe you can talk on Monday. How about that?"

"Sure," I say. "Hey, thanks a lot. Your help has been great."

He smiles and starts to close up his cabinets and desk. "No problem, man." He lowers his voice to somehow emphasize this point. As I turn to leave he calls after me.

"Hey college boy." I turn around and see him wink at me. "You should know. Jess isn't quite as nice and easygoing as me. He doesn't much like smart aleck college boys. You better step lightly around him." Dan is smiling wide now.

I'm not sure whether to believe him or not. I give a deep sigh. "Hey, thanks for the warning. Maybe I'll try to keep it a secret." I wave and wish him goodbye.

All the way back to my desk I am wondering what the problem could be here. I hope Vic will have some idea when he gets back Monday. I sure don't know enough about anything to make any guesses at all. Well, I think to myself, that is why you are a college boy. So you will someday know more than you do now. I am seeing value in getting an education already. And I am seeing some things I really enjoy about the engineering stuff. Anyway, I guess it is engineering. Thinking back on what President Hoover said, I decide that maybe it is the life for me. At least this S-line problem has become my phantasmagoria that is dogging my days and haunting my nights. Really, it's kind of fun.

Chapter 11

I am so pumped up about the charts and what they show, that I bring them to class with me to show Dr. Stromberg. I go to his office beforehand and show him what I have. He seems impressed with them, but points out that the chart of the 300 and 400 models are not really Pareto charts. He explains that Pareto charts show the greatest number of whatever you are measuring to least, starting from the left side of the chart. I remember that he explained that in class. He also says, however, that the way I had it might be a good way to easily compare the differences in the failure types from one model to the other.

After looking at the data for a little while he has a suggestion. "Drew, why don't you make two more charts? Here," he's pointing to the spring columns on each of the four charts, "look at these as you go from smallest model to largest. Notice anything?"

After looking at the totals I see what he is pointing to. "Yes, each failure count gets lower as the model gets larger. In fact, the largest one has hardly any spring failures."

"Correct," he says. "Also, there is a significant drop between the amounts for the 100 and 200 models and the 300 and 400 models. Do you see what I mean?" I nod. "That trend might be important in solving the problem."

"The other chart ought to be the same thing only with the cracked housings. There is a little bit of a trend there as well."

I see what he means. After looking at them for awhile I have to ask him. "Dr. Stromberg, how did you see those so quickly? I've been looking at this stuff for two or three days and working on the project for almost two weeks. How do you do it? Is there a trick?"

"No trick, Drew. Just getting used to seeing those kinds of things and naturally looking for trends and correlations like that, and knowing the tools to use to help you. In fact, let's just do a quick scatter plot. It is a plot to show if two variables correlate or have an impact on each other. Let's look at the total failures for each model."

He sketches a chart with model number, starting with 100, on the horizontal axis, then number of failure on the vertical axis. He has it done in just a few seconds. Then he draws a line from the top, by the mark for 100's spring failures to the bottom by the mark for the 400's spring failures. It is a pretty straight line.

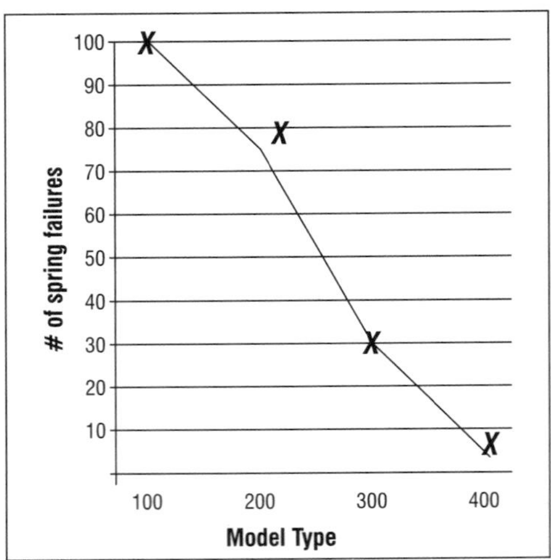

"This line represents the correlation of the model type to the number of failures. We call this a negative correlation because the slope is downward. This means as the model type gets larger, the number of failures gets smaller. A positive correlation would be when one variable increases so does the other."

"What if there is no correlation?" I ask.

"Then," he quickly sketches a chart on the back of the paper, "the data points are scattered, and you see no pattern at all. It would be nearly impossible to draw a straight line connecting the points. This means there is no relationship between the two variables you are measuring. Any questions?"

"No, I don't think so." This stuff seems like second nature to him and I get lost easily. It is pretty good how he came up with the neat chart. I think Vic will be impressed. I finally speak up again. "I'll go now. Thanks a lot for your help. See you in class."

He nods goodbye as I get up to leave, then adds. "Drew, if you do have questions, come and see me. This looks like a pretty good project you are working on. It would be a good choice for your semester project assignment."

I head to the classroom even though it is 15 minutes before class starts. No one is in the room so I find a good seat and, using the chart Dr. Stromberg did as a pattern, I do a similar one for the cracked housing failures. It looks similar to Dr. Stromberg's, except this one has two points at about the same level. Both the 300 and 400 models have 60 or so cracked housings. It looks like there is some correlation, even though the angle of the line is not as steep as with the springs.

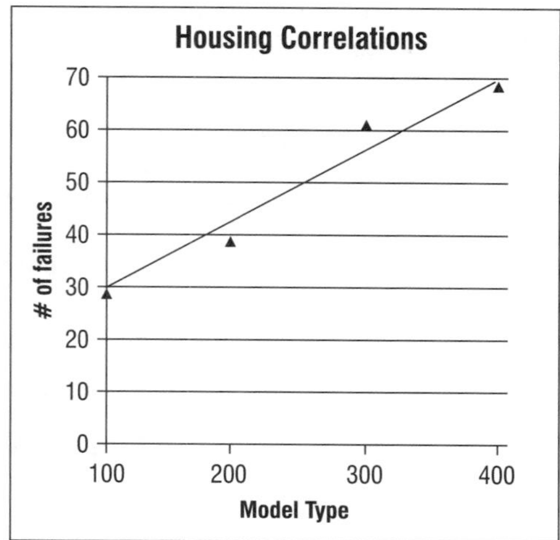

By this time others are coming in for class. Russell sits down by me. "What are you doing?" he asks.

"Just some charts for work. Dr. Stromberg showed me how to make something he called scatter plots. I'm just trying to make one of my own. You impressed?" I glance up at him as I show him the chart.

"Whoa, I'll say I am. Teacher's pet now, huh?" He teases.

I play along. "Hope so! Can't hurt the grade, you know."

By this time Dr. Stromberg has come in, and the class is quieting down in anticipation of class starting. It doesn't take too long and soon Dr. Stromberg begins class.

"In a recent class period we discussed the scientific method. Does anyone remember the steps of the method?" I hear pages turn as people check their notes. Soon someone raises their hand. Dr. Stromberg calls on them while adding, "I'm glad someone has kept good notes. What are they?" He asks the student.

"Hypothesis, experimentation, data and conclusion." The student looks up to Stromberg's approving glance.

"That is correct. And as I said, the scientific method is a basic structure of a process to gain information and understanding. It is a cycle of investigation through which we can learn things, and we can apply it to engineering principles and problems. In fact, it is upon this basic outline I want to build a problem solving process today. I hope to give you a couple of examples as well. Not every problem will fit the same way, but if used as a road map, this process will help you to logically and thoughtfully solve problems."

Now he puts the transparency on the overhead projector. It's similar to the figure he used last week showing the procedure of the scientific method.

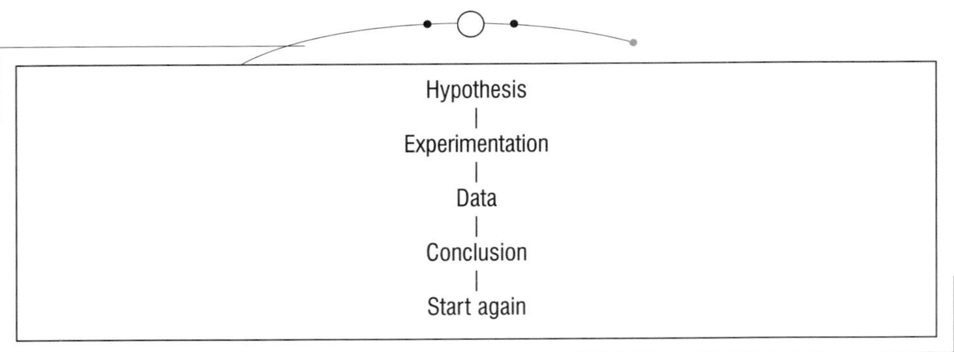

"Now I am going to put more common words and phrases on these steps and explain how to approach a typical problem." Dr. Stromberg already has my attention. I figure this might help me know how to proceed with the failure issue at work. I already know I am going to need a lot of help because I don't know the solenoid product well or what the materials are. Hopefully, when, or maybe I should say IF, we get a solution, I will be able to see how it happened. Stromberg has still been talking while I've been thinking.

"The second step is to try to define the problem in terms of the cause, not the symptoms. For example, if the car won't start, the problem is not that the car won't start, but that there is something that is causing the car not to start. The car not starting is the symptom. You want to search for the cause as to why it won't start."

I got step two, but didn't get one. Then I see it on the board and add it to my notes just above two. It is to "define the problem clearly and simply." Just as I am writing it down he adds, "The car doesn't start might be a good, simple problem statement, but we still want to strive to define the problem in terms of 'why.' In other words, by the cause, not the symptoms."

He is writing these on the board and starts the third point. Then he states out loud as he finishes the second sentence. "Gather good and valid information. DON'T DECIDE ANYTHING HERE." He adds emphasis as he reads it out loud. "The information might come from things right around the problem, such as other people, other machines, or the product. It might come from sources not close to the problem such as books you read on the subject, experiences you have had in the past, or others' experiences. Just make sure the information is," he points to the first sentence, "valid."

"Next, look for patterns, links, or trends in the data. These often tell you a lot or will lead to your answer. At least, it might give you hints to the answer." As I write this I glance over at the scatter plot Dr. Stromberg did for me before class. I remember he specifically mentioned the trend he saw in the data. This must be part of what he means. He gives an example and is moving on.

"Now after you have good information you try to 'determine the cause.' The next step is part of this." He starts writing the next step, "You must make sure to determine the CAUSE so you can 'design a solution that fits the cause, not the symptoms.'"

Now he steps back for a moment as if to give everyone a chance to catch up, pauses briefly until he sees that most are done writing, then continues.

"Lastly, implement and test the solution. Make sure you keep accurate and complete records so you are able to draw correct conclusions about the solution and make a determination from there as to what to do next."

He finishes the steps of his process then, underneath the list of steps, he writes one more thing, talking as he does. "Now, a piece of very important advice. Throughout the whole process be sure to think carefully. DON'T ACT FOR THE SAKE OF ACTING. Have a good reason to act." He again puts emphasis on the point.

After I have finished writing all the points, I look at the board to check and see if I have what he does. Dr. Stromberg is still at the board adding the four steps to the scientific method to the side of his points written on the board. I add the same to my notes and, when completed, I have copied what he has on the board.

Hypothesis

↓ 1) Define the problem clearly and simply.

Experimentation

↓ 2) Try to define the problem in terms of the cause not the symptoms.
3) Gather good & valid information. DON'T DECIDE ANYTHING HERE.

Data

↓ 4) Look for patterns, links, or trends in the data.

Conclusion

↓ 5) Determine the cause.
6) Design a solution that fits the cause, not the symptoms.
7) Implement and test the solution.

Start again

REMEMBER, always think carefully. DON'T ACT FOR THE SAKE OF ACTING.

After getting it all down, I sit back and look at it. This does look a little more understandable to me. I wasn't sure what those other things really meant, but I think I can do this. Dr. Stromberg is talking about an example of how to use these steps. I finally tune back in.

"Let's take a look at how this happens and use it in a simple problem you may come across in the course of your day.

"You come out of your apartment, unlock your bike and begin to take off on it. You feel a locking of the wheels, hear a grunt, and a clunk, and almost fall over the top of your bike. What happened? Assuming that you don't just take it to a shop and tell them, "fix it," but choose to figure out what happened, what are the steps that you may go through?

"First, you quickly, almost unconsciously, determine that since the bike stopped dead, it must be something that has somehow blocked a wheel. You don't begin with a check of the air pressure in the tires, because your experience or intuition sifts out the 'flat tire' reason almost immediately as the cause. The nature of the sounds and the feel and reaction of the bike tell you that you must look at something else."

He pauses here long enough to point to the first two steps he has on the board. He then adds, "notice we have already stated the problem simply — the bike stopped — and we are attempting to sift through the information to get to the cause, not just the symptom."

"So what do we do?" He puts up a handwritten overhead transparency.

- Why did I stop immediately?
- What were the sounds and what might have caused them? (Hint to the problem)
- Has this ever happened before?
- What part of the bike is likely involved that might be related to the problem?
- Do I have time to consider this right now? (Considered quickly but usually done)
- Have I done something recently that may have caused it to act like this?

I go ahead and write them down but I am having a hard time listening. My mind is starting to wander to the problem of the spring failures at work.

Stromberg continues talking, saying, "Let's take these questions one by one and determine if we can see how they come to mind almost naturally and help us form the problem and track down the ultimate solution. Though it might be very obvious and logical, we do ask ourselves, 'Why did I stop immediately?' This is a very important question and idea to consider, because we use it to begin an immediate brainstorming mode, such as ideas about what may be the cause. But it is also important because we make some immediate judgement about what the problem might be."

As Dr. Stromberg continues to explain his example, I am now caught up in how to try to use this at work. The next thing I know, people are closing their books and Dr. Stromberg is telling us to be sure and have the problems at the end of the next chapter ready for next week. I hope I didn't miss too much, but now I am going to get some lunch and head to work.

I am anxious to get to work, but when I get there I find that everybody I need is gone. Vic and Amy are still out of town. So much for sharing the great news with Vic. Dan and his friend Jess, of course, are duck hunting so I can't talk to them until Monday. Even Stan Hall was gone. Some executive retreat his secretary said, but she did say he had told her to expect me next week.

I ended up completing the two charts Dr. Stromberg had told me to do. They do show some interesting stuff. I hope I can learn to see stuff the way he does, and as quickly. After completing the charts I studied the CAD manual for awhile, then got on the computer and tried out a few things. It went well but my heart just wasn't in it. I ended up leaving work a half hour early to come back to my place and study. That is hard too. I just can't seem to get going on anything today.

I decide I've had enough of being in the room for awhile and head outside for a walk. It is a little cool but not too bad right now. Before too long that will change. I know this area gets colder than home and I'm not really excited about it. I will get to learn to ski, so it won't be too bad. As I cut through the parking lot, I see Stu sitting in front of his car looking at the headlights. "What's wrong?" I ask.

"Oh, these stupid headlights." I can tell he is not happy at all. "I'm supposed to be home in an hour for Mom's birthday party and the lights on the car have gone out. I was just going to the store to get new headlights to see if they would make a difference."

"Hey!" He looks at me with that 'you can fix my problem' look, and says, "You're the engineer. Why don't you fix it?" He sounds serious.

"I'm not an engineer." I feel myself taking a step back as if to tell him to keep me out of this situation. "I am just taking a class, and it's just an introduction."

Now he looks really down, not mad, just disappointed. Then I remember in our last class, Dr. Stromberg explained a process of problem solving that, he said would help us get to the bottom of any problem if we use it thoughtfully. Maybe it will help here. Besides, poor Stu could use the help. I decide to give it a try.

"Listen, let's first figure out what the problem is."

As I start towards the car, Stu gives me a look of hope. But as he hears me talk, it turns to a look of disgust. "I told you, the lights don't work."

"OK, right," I say, "the lights don't work, but before you rip out the lights and get new ones let's figure out what the root cause is."

I'm thinking now and wishing I had my class notes with me. First, according to Dr. Stromberg's method, you need to figure out if what appears to be the problem really is, or if it is just a symptom of the problem.

I continue. "Maybe it is the lights or maybe not. Sure, the headlights don't work but that may just be the symptom. Let's check into some other things. Does the car start?" I ask. Stu says that it will because he bought a new battery just a couple of days ago, but just to make sure he tries it again.

When he turns the key, the car immediately turns over and jumps to life. "Not a dead battery," I say to nobody in particularly.

Stu's getting impatient already. "The headlights must be burned out," he claims again. He won't let go of it.

"Chances are, Stu," I'm sounding pretty certain myself, "that both headlights wouldn't go out at the same time. I don't think it is the headlights themselves. Let's check a few other things." This is part of the next step that Dr. Stromberg described. Get as much good information as possible.

We check the radio. It works. The heater fan does as well. However, the overhead light does not come on and neither do the panel lights. The digital clock looks good and lights up when Stu turns the key. "This is crazy," I'm thinking to myself now, "some stuff works and some doesn't." I remember that Dr. Stromberg encouraged us to not jump to conclusions so I reserve judgement. The next step in the process is to look for patterns and common elements.

Then Stu's voice startles me. "What a piece of junk!" he cries. "The tail lights don't work either. Look at this." He points to the left tail light and sure enough, there's nothing. "There's no way to fix all these things. I better go call mom and tell her I'm not going to make it."

I stop him. "Wait Stu, let's think about this. Some stuff works and some doesn't." Even though more stuff doesn't work, I feel like we're making progress. "Look, the headlights, the tail lights, the inside overhead light, and check this out, even the park lights in the front. What's common about all of them?" I ask.

Looking at me as if I am nuts, Stu replies, "What's common is that they don't work, and what's obvious is you're acting like it's a good thing. I've got to go call Mom. If I don't leave in the next 10 minutes I'll be late anyway."

The thought hits me as Stu is talking, there is a pattern. I explain it to Stu. "Calm down Stu, what I mean is that all of the lights that don't work are the ones that usually do when the car is not turned on. Headlights, tail lights, inside overhead light, dashboard panel lights, all of them. Usually they work when the key is off and now they don't. None of them. The stuff that does work is the stuff that works only when the ignition switch is on, like the radio, wipers, that stuff. That means we have a pattern."

I am getting excited now. This idea of carefully thinking through a problem in a logical and careful way might just be something. Stu interrupts my excitement.

"You might have a pattern, but what I have here is a crazy man for a roommate. But I tell you what. If you can figure out what is wrong in the next ten minutes so

I can get to Mom's birthday party, I'll even take you with me and you can visit with Katie." I'm going for it. I wouldn't mind seeing Katie again to build a little bit of a friendship before she comes to SU next year. Could be a nice evening.

"OK," I'm on a roll now but I have to make sure I think straight. Dr. Stromberg taught us that one of the most common errors in problem solving is that people tend to become fixed on their first possible solution. What I think he meant is that too often we get anxious to get something done and don't carefully think through the problem. He called it "jumping to solutions."

"All the things that don't work must be on the same circuit. All we need to do is find what circuit that is and where the problem is. Pop the hood," I tell Stu. This is kind of fun I think, until I look under the hood. There are wires everywhere. How could one figure out this mess, especially in the 8 minutes I have left.?

"OK, Dr. Fixit," Stu pokes fun at me. "Where's the problem?" I start to panic. Wires everywhere and there could be a short in any one of them. It could be anywhere and it could be hidden. Well, I still have nothing to lose except a home cooked meal, some birthday cake with ice cream, and a chance to say hi to Katie. I remind myself to stay calm and think of a logical question.

"When was the last time the lights worked?" I ask Stu, ignoring his stabs.

"I don't know. I don't drive that much here. I know I drove it at night a week ago and they worked then."

I ask, "Have you done anything to the car since then?"

"What do I look like," he says as he looks at me unbelievingly, "Mr. Good Wrench? I don't do anything to this car but feed it and drive it. The only thing that has been done since I drove it a week ago is to have the new battery put in."

Dr. Stromberg mentioned that an important part of gathering good data is to make sure you know how the process, or whatever you are working on, has been behaving recently. I look at the battery. I've taken batteries out before and since we have plenty of power I don't expect anything. I'm hoping for some clue as much as anything.

I notice something that hasn't been apparent to me before. On the positive terminal there is the main power cable coming off that goes to the starter and switch. That makes sense to me. I also notice, however, a smaller wire that connects to the positive clamp. I think for a few seconds. If you have a wire coming from the positive terminal of the battery, it must be to get power directly to things without going through the ignition switch.

I trace the wire. Within a few inches I come to a connection where it splits into two separate wires that are all supposed to be held togther by a yellow plastic cone-shaped wire nut. I notice that one of the two wires is hanging loose from the connection. I know wire nuts are not the typical kind of connection used in car systems. I think to myself, "I wonder if . . . . naw, couldn't be that easy could it?"

Dr. Stromberg said that often complex sets of symptoms are rooted in a simple cause. I take off the wire nut, slightly twist the three bare ends of the wires togther and twist the wire nut back on. "Go try the lights" I tell Stu. He looks at me if I'm some kind of nut but obeys. He jumps in the car, hits the headlight switch and . . . . taa daa — light. Stu, of course, can't believe it. He looks at me, mouth open and speechless. I calmly look at my watch, "I believe I still have 3 minutes left. Shall we go?"

As we ride to Stu's house he can't stop raving about how brilliant I am. He's talking of starting our own company. With my brains and his brilliant marketing skills we could make millions, he says. I am pretty quiet. I am thinking about the steps Dr. Stromberg taught us. In class they sounded like just another lecture. Stuff to write in my notes and forget after the test. But tonight they really worked. I review them in my mind, modifying the words a little to fit what makes sense to me now.

Define the problem clearly and simply.
Try to define the problem in terms of the cause not the symptoms.
Gather good and valid information. DON'T DECIDE ANYTHING HERE.
Look for patterns, links, or trends in the data.
Determine the cause.
Design a solution that fits the cause not the symptoms.
Implement and test the solution.
and
All the way through, remember to think carefully. Don't act for the sake of acting.

It makes a lot of sense to me now. I also realize I have the example I need for my assignment on problem solving as well. This has been a great night.

When we get to Stu's house he makes me out to be the world's brightest man. Stu's dad shakes my hand and tells me he knows I'll be a great engineer. His mother gives me a hug and a kiss and thanks me for getting her son to her birthday party. I was hoping for the same from Katie, but she just says hi and gives me an admiring smile. I still can't figure out how Stu ended up with such a great sister. The evening goes well with good food, turkey and all the trimmings, plus birthday cake, good discussions with Stu's parents and Katie, and a happy time with the whole family. Like I said, it has been a great day.

Chapter 12

Finally I get to show Vic what I have come up with. Even though it's only 1:00, I am so anxious to get to work that I go in an hour early. As soon as I get to my desk I see Vic is there. He looks up as I come and glances at his watch.

"You're here pretty early, Drew." He smiles like he knows something as he says it. "Got something you can't wait to get to?" I'm not sure what he is being so smug about, but I pull the charts out of my desk anyway.

"Sure do, Vic. You were right about the springs. Look at these charts that show the S-line failures." I'm pretty excited and can't wait to show them to him. I know I'm talking pretty fast. "You're not going to believe this. It's really pretty amazing. I've been very anxious to show you this."

He laughs, "Slow down Drew. Actually I know you have some pretty good stuff. I happened to see Dan Johnson this morning. He was pretty pumped about the data you showed him last week. He told me this college boy I have working for me is a pretty good guy. I guess you really impressed him."

"So you know what these show?" I ask, a little disappointed I didn't get to surprise him first.

"No, I haven't seen the charts, I want to see them. Dan just talked about knowing that springs were the real problem. I thought the chart you developed had springs a close second to burnt contacts. What did you do, fudge some data?"

"No I didn't fudge any data, I just didn't count them right. I had grouped all the S-line models together. That was the chart you saw. But when I broke them out by model, the smaller models, the 100's and 200's, have serious spring breakage problems. Here look at this." I show him the Pareto charts that have the 100 and 200 failure problems on them. He looks intently then lets out a low whistle.

"I knew it was the springs. I knew it."

I stop him. "But look at this Vic." I show him the 300 and 400 charts.

His look of confidence changes to a look of puzzlement. "What in the world. . ." He stops for a moment then continues. "The larger sizes have hardly any problem. In fact their springs are really not a problem." He looks at me and asks, "Isn't this all the same material?" He answers himself. "Well, heck you wouldn't know."

I stop him. "Actually I do know. I asked Dan. It is the same material. And all the springs are made in-house. No outside vendors. Annndd," I'm having fun now, "it's a new material."

Vic looks surprised. "Been doing your homework huh, college boy." (He must have heard Dan refer to me like that.) "So then I guess you know what the problem is then?"

I stop my cocky talk and have to admit I don't. "But Dan says he has a hunting buddy in the spring shop that might have an idea. I'm going to go talk with him in a while."

Vic gets more serious now. "Well done, Drew. That would be a good idea to talk to Dan's friend. I think I will talk to Jack Danvers, he's the engineer that worked on these models originally. I think I will also check with Carl Sanderson. He's the manufacturing engineer that worked most of the processes for the solenoid parts. He's has a pretty good materials background also."

He stops talking for a minute then asks me, "How did you come up with the idea to do separate charts? That was a good idea. I'm not sure I would have thought of it since all the springs are made out of the same material."

I'm feeling pretty proud he likes this stuff. "Well, I didn't at first. Then when I was talking to Dan, he happened to mention that he knew the smaller models had more springs breaking than the large ones. He was pretty confident. Then it just so happened that we had just had a lecture by Dr. Stromberg in my intro class about charting. Dr. Stromberg taught us about Pareto charts and told us to look at charting data in different logical ways. He said to look for natural divisions and ways to separate data. After talking to Dan, failure type by model seemed logical. That's how it happened."

He nods his head as if to agree. "Good engineering thinking Drew. By the way, keep listening in class. You will learn a lot. One of the unfortunate things about education is you usually learn a lot of good stuff you don't have any use for at the time. You are lucky that you have something to help you use what you are learning. It's good for you and," he points to the four failure charts, "it is good for us."

He stands looking at the charts for awhile longer. I can tell he wants to say something else. What's really neat is I know I have done something that Vic really appreciates and it feels good. But even more than that I can tell that Vic wants to help me learn something more from it. Finally he says something.

"Drew, keep a notebook of what you are learning about these things. Pretty soon all this information you are obtaining will seem like it flows together. If you don't have good documentation you will forget things. Here, look at this." He picks up a notebook from his desk. "I keep track of meeting notes, visits with others in the plant about work issues, ideas and thoughts that come to mind, all those kinds of things in here. Make sure you put down the date and time of your entry and who you visited with. Look at this example." He points to the top of the page.

September 13, 2:00 p.m.
Metal Pro visit regarding new body material for the S-line series.
Attending: Vic, Amy, and John from ACTC.
Shelly Daniels and Robert Torner from Metalpro.

Notes: The stress tests on the performance of the body are back with good results. Failures occur at pressure levels 20% above the highest we have had. The tests look good. They will have leak tests back after next week on all four sizes.

I think we better have them review their alloy compositions with our metals people just to be sure.

"See here? I have indicated who we talked with and who was there. The day and time and a few notes about the visit. This note about having our metals people check on things is a thought I had on the way back and wrote down after the meeting." He closes the book and tells me if I need a notebook there are some in the supply cabinet.

"OK, back to the issue at hand. Next step, you go talk to Dan's friend in the spring shop. He may have the experience that will help us. I'll see the other engineers and get some information on the material and process theory. Theory and practice together, Drew. That's the key."

He turns to walk out. On his way out he squares his arm and points his finger in the air like someone with something profound to say. In a deep voice he says, "Remember. In theory, theory and practice are the same thing. In practice they are not." With that he is gone. I'm not quite sure what he meant, but it gradually sinks in.

I pick up my charts from where Vic left them and head out to the test area. As I'm walking I think about what Vic said about notes. I haven't been writing anything down to this point. I guess I kinda did figure I would remember it all. It is a pretty good idea but it seems to take too much time. I'm not sure what to write anyway.

I remember a lecture we had in the intro class a few days ago. Dr. Stromberg talked about documentation and keeping what I think he called work notebooks.

He gave the example of Thomas Edison. He said Edison kept very good notes and had hundreds of work journals. Unfortunately, like Vic just said, I didn't really see much application for the lecture, so I didn't keep very good notes. I think Vic just gave me a reason. Just then I stop walking.

"Oh man," I hit myself on the forehead as I mumble. "Here I am thinking I better take notes, and I have nothing to write on." I almost decide that I'll start later, but think better of it and head back to the supply cabinet to get a notebook. On the way I keep thinking about the stuff Vic told me. First, keep notes of what I learn here. Second, try to pay attention in my classes because chances are I will need this stuff sometime. It is almost like having faith. Third, theory and practice are both important. This one isn't very clear to me yet but appears to be something that Vic believes based on the way he said it. Just then I realize that I am next to the supply cabinet. I open it and see lots of loose leaf paper and 3 ring binders. I start to take some, then I notice a stack of bound notebooks. I think these will be better because there is less chance of losing important sheets.

Finally I'm back to the shop with my new work notebook, charts in hand and ready to learn more. I see someone standing at the test bench with his back to me as I enter the testing shop.

"Hey Dan." I call out and he turns around, but it's not Dan.

"He's not here." The guy who was standing there is working on a solenoid.

"Sorry" I tell him. "I assumed you were him. Do you know where he is?"

"Yep. Took off at noon to go hunting. Won't be back until tomorrow. I guess he must be having fun or he got stuck. What do ya need?"

"Oh he was helping me on some failure stuff for the S-line solenoids. I guess . . ."

The guy interrupts me. "Oh yeah, he told me about that. He said he and some college boy got some problem fixed. You the college boy?"

I sure wish people would stop calling me that, but I nod my head. "I guess I'll talk to him tomorrow. You sure he will be back then?"

"Oh I think so. A man's got to work sometime if he wants to keep the bills paid. Dan said you and he were going to talk to Jess over in the spring shop. Jess is with him so he should be back tomorrow too. Sorry." He shrugs his shoulders.

"It's OK. I'll catch him then. Thanks." If people don't stick around I may never get this done. Oh well, guess I'll head back and practice some CAD stuff if nothing else.

I spend most of the rest of the day correcting a few things on the charts and printing out new copies. About 4:30 Vic comes in.

"Hey Drew. I couldn't find Jack, you know, the design engineer, but I did talk to Carl, the guy who really understands materials. I told him about the numbers and he would really like to have a copy of your charts. Copy a set for him will you? Before you do that though, make sure your name is on the bottom of them."

The confused look on my face makes him explain. "Listen, two reasons for that. First, those are your charts and if somebody has a question or wants some explanation, they ought to know who to contact. Second, that is also a way to make sure you get some credit for the work you did. It's good stuff and you deserve some credit."

Makes sense but it is kind of scary. "I don't know much, Vic, and besides I'm not here most of the day. Maybe I should put your name on it too."

He laughs. "OK, maybe that would be good. Listen, just put on the bottom. 'For more information contact,' then put your name and date. Sound OK?"

I nod my head and pull up the file on the computer and make the change. I make copies and leave them on Vic's desk. By the time I get all this done it's almost five. I wrap things up and head home.

As I'm leaving, out of the blue I wonder how Mom and Dad and the others are doing at home. I have no idea what brought it on. I've been here at school for just over a month now and it's been good. Am I homesick? I don't know, but I think maybe I'll call home tonight. I've been wondering how Dad is. I know he won't know much more about his prognosis until Thanksgiving, but it'll still be good to talk to him. It's been almost two weeks since I talked with them. I'm still thinking about what Vic said today about listening in class because you'll need it sometime. I know I have already seen a bunch of things that Mom and Dad kept telling me at home that I think I appreciate more now. Maybe it applies to life in general. I got along OK with Mom and Dad but we had our times of disagreement. Maybe what Vic said applies to stuff learned at home too. I'll bet he knew that. Man, this is deep stuff. I better back off. Could I be growing up? Boy, that thought would shock Mom and Dad.

I get to the bus stop just in time to catch my bus. It's been another good day. Can't wait for more.

Chapter 13

The next day after classes I head out early to work again. I make it in about 1:15. Vic looks up from his desk as I walk in. "You have a message on your desk that was dropped off about 5 minutes ago. Looks pretty important."

I go over and pick it up.

> Drew:
> Sorry I missed you last week when you came by. Stop by sometime this afternoon if you have a moment. I'd like to chat with you.
> Thanks, Stan Hall.

After I finish reading it Vic looks up at me. "Still hobnobbing with the big shots huh? I wouldn't keep a VP waiting if I were you." He gives me a wink as he goes back to his work. I'm not sure what to say so I just tell Vic I'll be back in a few minutes and head toward Stan Hall's office. I feel a little nervous but excited. I know it is a good chance to learn, and besides what a neat thing that the VP knows who I am. I still feel a little silly when I think about what happened at the front doors my first day here. Of course, if Stan Hall hadn't come along when he did, I probably would not be working here. I felt pretty intimidated.

As I walk into the secretary's office, she looks up and seems to recognize me. She invites me to have a seat and goes into Mr. Hall's office. She comes out pretty quickly to tell me he will be with me as soon as he gets off the phone. I thank her and relax a little. Around the secretary's office there are a number of framed photographs of different machines and parts. There are planes and cars and some stuff I don't really recognize. Each one has a caption plate with a short sentence describing what ACTC part is used and what the machine is. The one under the airplane says "ACTC S-line solenoids make Baldwin's planes fly."

There are two paintings on the wall, one behind the secretary's desk and one behind the guest sofa, that are not of ACTC products but are just paintings. "What are these?" I ask the secretary.

She smiles. "Mr. Hall appreciates good art too. These are a couple of nature scenes from one of his favorite artists. Do you like them?"

I nod, but I am a little surprised. "Yes, they're nice. I didn't know Mr. Hall was a fan of art. Does he have more?"

Just then Stan Hall comes out of his office. "Hello, Drew. Good to see you. Nancy told me you had stopped by last week. Sorry I missed you."

Now I feel embarrassed again. "Oh that's fine. I had a few minutes and just came by to see if you were here. You had told me to come this week. Sorry."

He puts his hand on my shoulder as if to guide me to his office. "No matter now. You're here and we can talk a little."

Nancy points to the art print behind her desk. "Drew was just asking about your prints. He seemed a little surprised you enjoyed art."

Stan laughs and says. "Well maybe we will talk about that a little bit, too. Come in, Drew."

It is a spacious office but not too big. I mean big enough to seat a number of people around a small conference table and still have plenty of room for a desk and a couple of guest chairs. There is a beautiful picture window with a view of a pond, small waterfall and trees lining the pond's edge. There are a number of ducks swimming in the pond. A small path leads through the trees and along the edge of the pond with a couple of benches for resting.

"I didn't know this was here." My surprise is obvious to Stan Hall. He is standing behind me.

"Quite pretty, don't you think?" I nod and he continues. "This pond is fed by water from our plant. You may or may not know that we use a fair amount of water in our processes, and it needs to be treated to remove any contaminants and heavy metals. We then filter it again and direct it into this pond via that waterfall over there." He points to the waterfall that first caught my attention.

"If the grass won't grow, and the fish can't live, and the ducks won't stay in the water, then we must do a better job. We have a responsibility for the things around us, don't you think, Drew?"

I nod but must still have a look of bewilderment on my face. "Yes sir, it really is beautiful."

He smiles. "ACTC has always placed a lot of value on the world around us and our responsibility for taking care of it, as well as the community. Engineering is, after all, a profession dedicated to improving the way we all live."

He goes over to one of the guest chairs and sits down. He motions for me to sit as well. "Drew, how are things going here? Are you learning?"

"Oh man, I sure am. This has been great. I haven't even been here a month, and I feel like I have learned a ton. It is really great."

"Vic tells me you are making some good progress on the spring problem on the S-line series. He says you have done some very good work."

I flush a little. "Well, I really haven't done too much except develop a few charts and talk to some people about the problem. Vic has been a lot of help, and Dan Johnson in the test shop has really helped, too. I think we have learned some things but all I've really done is the charts."

"You will find, Drew, that good engineering work, for that matter any work, involves a lot of people. Talking to others and learning what they have to say, getting their view of things, is critical. Keep that up. I suspect you do it now because you feel you have no answers by yourself."

"I don't know much. As you know I just barely started school."

He continues, "Well that's the way it should be, and it should continue to be that way regardless of how much you think you know. No matter how much you know, you won't know enough to solve most problems by yourself."

He gets up and walks over to the window and looks out. "In fact, Drew, that is part of the reason I wanted to talk to you. Not just to encourage you to include others in your work. It looks like you are learning that just fine. But to encourage you to gain different perspectives from other sources as well."

He comes back over and sits down. "What classes are you taking right now?"

"Well, I have the Introduction to Engineering class, a freshman English class, a math class, an art history class, and a physics class. Let's see, that's 16 hours, isn't it?" I count them up quickly in my mind. "No that's 15, but I also have a PE health class, that makes 16 hours."

He nods as if to agree. "That is a pretty good load while you are working, too. How is school going?"

"Pretty well. I'm not failing any classes, but I guess I could do better in some of them."

"Good," he replies. "Now a couple of those are general education classes aren't they?"

"Yes, the English is. But that is also required for all the engineering majors. Then the art history fills a GE credit. I thought it would be good to take some and get them out of the way."

This brings a smile to his face. "That's the same thing you said the day I first met you. Why do you say, 'Get them out of the way?'"

"Oh, I don't know. I guess that is what everybody says, and it seems most people feel like the GE classes are not quite as important as the major classes and just kind of get in the way. I guess they really don't, but that's kind of the view." I wonder why he is asking. I now remember the first day I met him he said something about coming to see him about GE classes. I thought maybe he was going to give me a way around some of them, but I get the feeling that's not the case.

He sits forward a little in his chair. "Drew, you're right. Many students do see it that way, and not just in engineering. It is also seen that way in many other disciplines. I don't think it should be. You are at school to become educated. Education means more than learning just a field of study or preparing for a career. Those are important, but an education is the opportunity to gain learning about theories, history, principles, and experiences of many different things. Classes you take in school in all kinds of fields and disciplines are valuable to help you see the physical, social, and historical world much better. This will make you a more valuable citizen in the world, and by the way, a much better engineer."

He walks over to a small figure statue on a table near his desk. "So do you know what this is?" he asks me.

I do recognize it from an art history class I took in high school. "I know it's a famous statue, I think it is called *The Thinker.*"

He nods his approval. "That's right. And that's what too many people don't know how to do. Think. Your GE classes will help you become a better thinker. Learn to appreciate them. I know that college can be a bit of a hassle at times. The grading system, the disintegrated way things are taught, the way too many professors worry more about their research and publications than they do about your learning. But in spite of all that, it is the best system in the world, particularly if you take advantage of it. No matter what the professor may do in any class, you can take advantage of the class materials to become a better learner and a better person."

He stops talking and walks back over to the chair. He sits down and looks at me. He smiles and says, "Wow, guess I really got off on preaching that sermon didn't I?"

I smile back. "Thanks, though. I guess I never really thought about the advantage that the GE classes can be to me. I guess I knew they must be important somehow, but I think I saw them more as a hoop to jump through for graduation than an opportunity. I'll see them differently now."

"Good," he replies. "Remember, we are in a world full of all kinds of information. Information is good but it can't make decisions without some intelligent interpretation. Information leads to knowledge, which leads to understanding, which in turn leads to wisdom. I like the increase on thinking ability suggested by that sequence. The broad range of subjects you can learn about, even outside of classes, will give you the perspective to be a better decision maker. Art, for example, as you can tell I enjoy, is best appreciated as you view it from different perspectives. My ability to solve problems has been positively influenced by my learning about art."

We chat for a few minutes more about the specific classes I have, then I excuse myself and head back to my desk. I really do feel grateful he took time to explain his view to me. It makes sense. The only problem is I may have to stay awake more in my art history class.

When I get back Vic is still working at his desk on some part prints. He looks up as I come in and teases, "So are you my boss yet?"

"No Vic," I answer, "I am not your boss yet. Stan said I would have to wait until after I finish my Intro to Engineering class, and if I pass, then I can be your boss."

He laughs then says, "Hey Drew, come here a minute, I need your help."

I'm not sure what I can help with but I go over to his desk. He has a chart of some sort on his desk and is looking at it quite intently.

"Drew, this is a Gantt chart for the housing redesign."

He sees my confused look. I ask, "A Gantt chart?"

"Yes, it's like a timeline of tasks that need to be done and indicates when they will get done. I'd like you to help on the layout and detailed drawings for the redesign. That's why I've had you learning the CAD package. By the way, how's that coming?" He waits for an answer.

"OK. Actually, I haven't concentrated on it as much as I should because I'm so anxious to get the springs figured out." It's the truth and now I feel badly that I haven't done more.

"I understand," he says. "But when you can, put some more time into it. I'll get one of the designers to help you, too. That will speed your learning curve." Fortunately we have had some lab assignments in the Intro to Engineering class I am taking that teach the basics of CAD. It has been a great help to learning because even though ACTC doesn't have the same system as the school does, the basic principles are the same.

He turns his attention back to the chart. "What I need your help on is this. I want to make sure you will have time to get involved in this project as we get close to the layout and detailed design tasks. Here, look at this."

He shows me the chart. It has a series of tasks down the left side, upside down triangles that indicate weekly time period, and horizontal bars that look to me as if they show how long a task will take. Vic lets me look for a moment, then speaks again.

"We have already started on the first and second tasks. As you can see, by the end of the forth week, which is the week after Thanksgiving, we are to have finished developing the alternatives. That means we need to start documenting the design," he points to the task labeled 'layout design' then continues, "by mid-December. Do you think you'll be done with the spring project, or at least far enough along you can help with the design?"

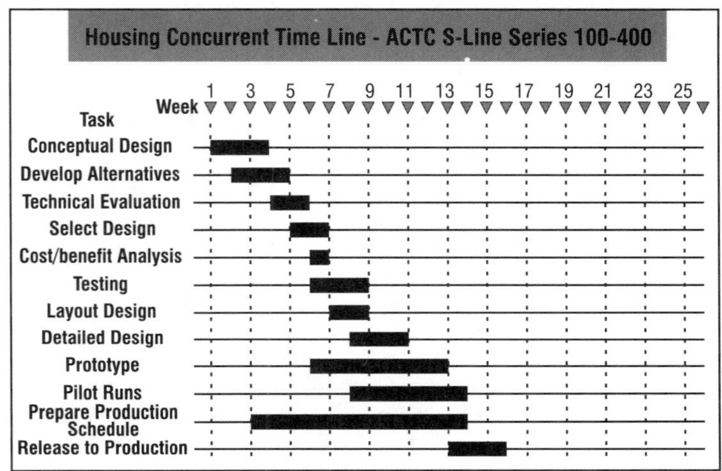

I'm not sure how long anything will take and I'm afraid to make a commitment. "I'm not sure Vic. I'll make sure I spend more time on the CAD system, but you're probably better at estimating when we'll be done with this project."

"Well," Vic responds after thinking a moment, "I think we'll be wrapping it up by then. I'm going to leave you on the plan to help out. Is that OK with you?"

"Sure," I tell him, but now this chart has piqued my interest. "But can I ask you another question?"

"Sure, what is it?" He's writing my name beside the design tasks.

I point to the task labeled "Prepare production schedule." "How do you do this when you haven't yet finished the designs? In fact, there's a lot of overlapping tasks here that seem to me need other tasks completed before they can be done themselves."

"Very observant, Drew. Let me explain something to you. Notice this." He's pointing to the word "concurrent" in the title. "Time to market is a very important concept. Time to market is the amount of time it takes us to start a project until the product is in our customers' hands. To stay competitive, we must be faster than we used to be. So we do overlap tasks, but we actually get a better product that way. Here. Look at this Gantt chart." He pulls another chart from the stack of papers on his desk.

"See here." He keeps talking as he points to the word "sequence" in the title of this chart. "What I did first was lay out all the tasks and assign the times each would require to complete. That's what this chart shows. Look at the time to compete the whole project."

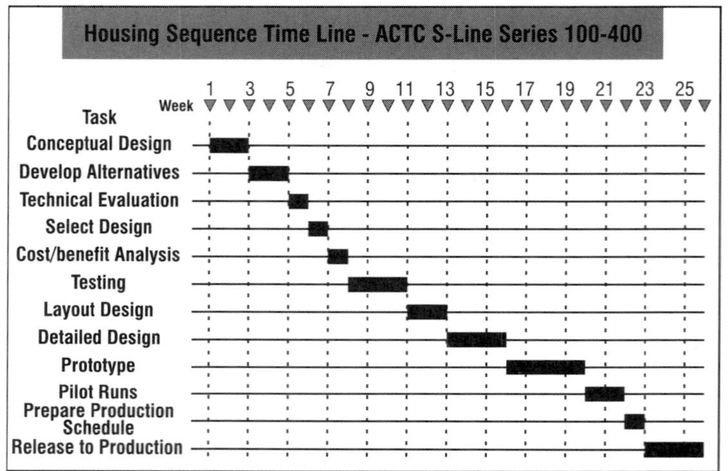

I look at the chart and say quietly, but loud enough that he can hear, "About twenty-six weeks."

"Correct. Now look at this one." He points to the first chart, the one marked 'concurrent.'

"Sixteen weeks. Wow, that's two and a half months sooner." I hold the charts side by side as I look at them.

"That's right, Drew. Now that's two and a half months of profit, and hopefully an edge on the market. That's very important as you might expect. We call this," he says, pointing to the concurrent chart, "concurrent engineering, or some call it simultaneous engineering. It's the idea that we get people working on tasks concurrently to reduce the time necessary to complete the whole project, and we get people talking together about things so they can make better decisions."

He points to the task I first looked at labeled Prepare Production Schedule. "From my sequence chart you can see that this task really only takes 1 week to complete. So when I put it on the concurrent chart, I'm not increasing the amount of time, only the moving start time. We will have production planners involved early in the project. In fact, I'm going to see them tomorrow, so when we get to the point of needing a schedule it will be nearly done. As you can see the same is true of testing, prototyping, and the pilot runs."

I'm still not sure how this works. "Don't people get confused, or forget what they are working on?"

"Not any more than before. In fact, we plan and track our projects better this way and that helps reduce the confusion. Here, I want to point out one other thing. See this?" He's pointing to the time for the first two tasks. "We spend more time planning at the beginning of the project now. This helps anticipate and resolve concerns that used to come up later. This whole idea of concurrent engineering has helped us a lot. Better products in shorter time."

It looks pretty interesting. I have a couple more questions. "So this project is already underway. What are you doing now? Are you changing something?"

"Not really. I'm reviewing the tasks to make sure we are on track and to plan resources. I try to do that a couple of times a week to make sure we are following the plan and, if I'm lucky, shorten the project by a week or two. Any other questions?"

I shake my head. "No, but thanks. That's pretty interesting. Maybe I'll learn more when I get involved in the project."

Vic smiles. "I'm sure you will."

I leave Vic's office and get the charts out of my desk, and the notebook I have started to take notes in, and head for the shop to find Dan. I want to talk to him about these failure reports and have him introduce me to Jess in the spring shop.

When I get to the test lab, Dan is there but pretty busy on some stuff he says he has to get out today. He says that taking a day and a half off for hunting put him behind. He also says he has talked to Jess and he will be expecting me. He gives me directions to the spring shop and sends me on my way. I was hoping to have Dan take me over. I get a little nervous meeting people for the first time. I feel like I'm too young to be doing this stuff, but it's got to be done, so I head out in the direction Dan pointed me.

The spring shop, though part of the company, is in a separate building. I follow the directions and find the building, but I can't find Jess. Finally I ask somebody where he is. I am pointed toward a set of machines on the far side of the building. I finally find Jess standing beside one of the three machines in the area. He looks busy so I just watch for a few minutes.

Close by there are fairly large coils of steel wire of various diameters. These must be the raw stock for the springs. Some of them have diameters as large as a toothpick while the smallest are about the size of a hair. The stock feeds into one of the three machines through a hole in the back. Then the wire is wrapped around a solid metal bar and cut, and then a hollow barrel removes the newly formed spring which drops out into a bin. It repeats the process with each machine working relentlessly.

Jess looks up and sees me. He motions for me to come closer. "What do ya need?" He talks loud to be heard over the drone of the machines. "You the college boy Dan was talking about?"

I nod my head. "I guess I should change my name to 'college boy'. Seems I get called that more now than Drew," I answer him. "I'm the one Dan told you about."

He puts his hand up to his ear as if to tell me to speak up. I repeat what I said only louder. "Yes, I'm the one Dan told you about. We thought you might have some ideas on why the smaller springs for the S-line series of solenoids break more often than the bigger springs."

He nods then puts his finger up and says, "Hold on a minute, I'll be right with you." He checks some settings on all three machines, looks at the feeder stock then waves me to a little desk a few feet away.

He doesn't need to talk quite so loud here. "I'm not sure what I can do for you. I've been doing these springs for about 8 years now so I know how they're made, but I don't know much about the stuff behind them like the engineers do. What kind of questions do you have?"

Jess is much different from Dan. He's more soft spoken and seems a little subdued. Dan can make you feel like you were being bowled over. Jess is older, I'd guess about 50 or so, with a slight build and about two inches shorter than me.

"Well," I try to think of something intelligent to ask him, "I guess Dan told you that a lot more of the smaller springs come back broken than the large ones. Any idea what could be the cause?"

He has a thoughtful look on his face and doesn't say anything for a while. After some pondering he shakes his head back and forth. "I really don't know. We do spot checks on the lots for all the spring sizes," he says as he points to a chart on the wall right by where we are standing. "And on the tests we don't see any higher number of the small ones breaking than we do the large ones. I'm afraid I can't help you on that."

He tells me to hold on a minute and he goes over and checks some stuff on the machines. He makes a couple of adjustments and then returns.

I ask another question. "You said you've been here for 8 years. That means you saw the change to the new material. Did you notice anything then?"

He speaks up right away. "Actually, not too much changed. We are using the same machines—well, at least the same kind. We did get a new one that is a little more automated, and we hope to replace another one soon, but the machines and the process are pretty much the same. What did change was the amount of breakage on our sample tests. When we started using Springalloy the breakage went way down. Good stuff that Springalloy. Other than that, nothing much has changed."

I repeat his answer in the form of a question. "So nothing really has changed other than a new machine and the material, huh?"

He moves his head back and forth, then stops. "Well one thing, but I don't know if it is a big deal or not. We used to cold wrap all our springs. Now we hot wrap them. The specs on this new material call for the material to be heated before we wrap them. That's what those long tubes are on the back of each machine—ovens to bring the wire up to the proper temperature. Other than that, it's the same right down to the mandrels."

I write that down in my lab book along with the other notes I have taken. I thank him and turn to leave. "Hey," he calls after me, "you going into engineering?"

"I'm studying it right now at SU, but I'm new there and haven't really decided what to do yet," I tell him.

"Good for you. My boy went to SU in engineering. He works up in Lake City. Has a great job and really seems happy. Stick with it and you'll do well. It's pretty good money, and challenging work, too. Good luck, and let me know if I can help." He waves as he walks back to the machines.

It takes me almost ten minutes to walk back to my desk. All the way I'm wondering what to do next. Seems to me I'm at a dead stop. No leads in the case. I really had hoped to get a breakthrough with Jess like I did with Dan the first time I talked to him. When I get back Vic and Amy are talking about my charts.

"Hey, just the man we wanted to see. What did you learn from Jess?" Vic waves me into Amy's office.

"Nothing really. I'm not sure what to do now. You have any ideas?"

He shakes his head. "Nope, but Amy thinks we ought to sit down with Carl and go over what we have. You OK with that?"

"Sure. I don't know what to do now anyway. I don't know anything about the material or much about the processes. I wish I did. When can we meet with Carl?"

Amy speaks up now. "Carl is out of town until Friday, but he said he could meet then. We will meet Friday afternoon right after you get here, Drew. Is that OK with you, Vic?"

We all agree. Actually I'm disappointed. I would rather meet sooner, but I know the others have a lot more to do. So Friday it is.

Chapter 14

On Tuesday nights the foreign language department presents selected international films. Tonight they're showing a new Italian film. Even though it is not art in terms of what we usually call art, I can get credit in art history for attending a foreign film, so I decide to go. I called Christy and she was willing to go on short notice. I pick her up at 7:00 so we can get to the film by 7:30. It's about a 15 minute walk from her dorm to the building where the film is being shown. We have a nice chat on the way about nothing in particular. She says school is going well for her but she still doesn't know what she is going to major in. She thinks she still wants to study education, but is also thinking about business. She thinks it may be easier to make decent money and might have more opportunities.

She knows I am working at ACTC and asks how it's going. I tell her it's OK. I don't get into any details because the last time I did I couldn't stop talking about it. She seems very interested in what I'm doing. We start talking about the classes I'm taking, especially Introduction to Engineering and Engineering Technology. She is even more interested. The more we talk, the more I realize this might be the right kind of major for her. I know that she's bright and a good worker. She told me she worked during the summer for her dad making wood cabinets. She enjoyed it, and it sounds like she did well. The opportunities for women in engineering and engineering technology are very good. I tell her about Amy at work and how well she is respected. She wonders out loud if she could talk to Amy. I am sure Amy would love to, but I tell Christy I'll ask to be sure. By this time we are at the auditorium and the movie is starting so we stop talking.

The film isn't too bad. It's Italian with English subtitles. It's about a lower class Italian guy who gets a job as a postman in a rural town. As part of his route he delivers to a well-known poet who has a nice home near the town. The postman gets to know the poet. He also has a crush on a local barmaid, and he tries to attract her by writing poetry. The poet helps the postman with poetic metaphors. In the end, the poor postman dies in a revolution that is taking place that also forces the poet to leave the country. So, the postman dies and doesn't get the girl, the poet leaves but he does come back to pay some honor to the postman, and the movie ends. It was OK, but not much action.

As we are walking out Christy says, "Wasn't that romantic?" She really seems to have liked it. I guess I kind of look at her funny, because she stops walking and declares, "Didn't you see the meaning of the movie?"

I'm not sure what to say but try to respond. "Uh yeah, don't deliver mail to a poet or you could end up getting killed." As soon as I say it, I realize I've made a mistake but it's too late.

"I can't believe you. Are you kidding?" She looks shocked. I try to recover because she gave me an opening.

"Well, ah, yes, I am. What I mean is, it has some pretty cool views of the ocean and the postman guy tries to write a nice note to the barmaid so she would like him. Good stuff." I hope my recovery worked, but one look at Christy's face tells me that it didn't.

"You're joking right?" She waits a couple of seconds for me to answer, but I don't want to get in any deeper trouble. Then she continues, "Don't you see that the whole movie is a metaphor? It's a symbol of the power of love and what a person will do for love. The postman's undying devotion to the serving girl, his learning to write to win her love, and finally, giving his life to defend his belief in her and the poet. Wasn't it wonderful how he changed from a lower class peasant to a poet himself, expressing metaphors, symbols, and words that caress the soul?"

If she hadn't been sitting beside me the whole time, I would have wondered if we saw the same movie. Trying to be real careful now, I just say, "Oh yeah, I see what you mean, it really was great, wasn't it?" I decide a diversion might work at this point. "Christy, would you like to stop for a soda?" She looks a little frustrated but says yes, and we head into the student building.

The rest of the time we just chat about school. She brings up the movie a couple of times. I listen and agree with her. Actually she has some pretty good thoughts on it. Stuff I didn't notice. We finish our drinks, and I walk her back to her apartment. She thanks me, and I head home.

When I walk in Stu is there already. Not only that, but he appears to be reading a book.

"What are you doing?" I look at him in surprise.

He looks back and answers. "Studying." I stand with my mouth open and a look of surprise on my face. He laughs and says. "Hey don't be so surprised. I do study regularly, you know. This is for this month, and I did once last month, too." He then asks, "Where you been?"

I plop on my bed, the kind of plop Mom always gets mad at me for, "I have been to a metaphor of love and devotion as symbolized by a common postman giving of himself to win the love of a barmaid . . . no, I mean serving girl."

Stu starts laughing and calls me romantic. Maybe I am learning. Actually in the long run, it worked out OK. Two days later, in my art history class, the teacher asked if anyone had gone to see the movie. When I raised my hand she called on me to tell what I thought. I stammered for a minute then basically repeated what Christy had said. That the whole movie is a metaphor. That it's a symbol of the power of love and what one will do for love. That the postman had undying devotion to the serving girl and did all he could to learn how to write so he could win her love, even giving his life to defend his belief in her and the poet. Wasn't it wonderful how he changed from a lower class peasant to a poet himself, expressing metaphors, symbols, and words that caress the soul?"

When I stopped the professor looked at me suspiciously. "Well, very interesting, where did you get that interpretation?"

"Oh, I got it from the movie. Good show." I decide to say no more.

The professor still had a suspicious smile and the girl in the row next to me smiled and said, "I'm impressed." I smiled back, wisely keeping my mouth closed. The professor was still wondering. I left it at that, realizing I came out OK. Maybe

I can learn something about how others view things, but I still wonder how the postman metaphor is going to help me solve the S-line material problem.

Friday finally comes. On Wednesday and Thursday I spent most of the time learning the CAD system. Between the labs at school and the on-line tutorials I feel like I am doing pretty well. After you get used to the commands it is really kinda fun. Vic wants me to be prepared to help him do some design changes for the new S-line model upgrades they are planning.

In the conference room are Amy, Vic, Carl and myself. Just as I sit down I hear a voice behind me.

"Hey there, college boy." I turn around to see that Dan has just walked in.

I'm really glad to see him. He's been an important part of this. "Hello, Dan, good to see you here. Is Jess coming too?"

Vic interrupts before Dan can answer. "Jess can't come today. He has a big order of springs that need to be done today for transfer to assembly by Monday. He said Dan would represent him."

Dan jumps in. "Well college boy, I don't usually take the time to attend these big time meetings, but since I was the one who put you on the right path on this problem, I thought I'd come along and help out."

Amy speaks up, "Let's get started." And everyone sits down. "As you all know we are here to talk about some of the work that Vic and Drew have been doing on the types of failures we have had on the S-line series. Vic, do you want to describe what you have here?"

"Sure, I'll tell you a little bit, but I think it will be best to have Drew and Dan tell you about the actual counts. I had Drew take a count of all the failure types on the S-line series because I have thought for a long time that the springs are a serious problem. As it turns out they are, but not quite like I had thought. Well, listen Drew, why don't you explain what we have here?"

I hadn't expected to say much, and I feel pretty intimidated in this group. I brought copies for everyone so I decide to start there. "Here are copies of the charts we have so far. They show that the springs are a serious problem but mostly on the 100 and 200 models. The large solenoids really don't have many spring failures."

Carl interrupts, "How did you decide to do this kind of breakdown?"

"Well, actually my professor at school told me about this type of chart. When I was talking to Dan one day mentioned that the problem was on the smaller models and not the larger ones. I just took the data and showed the failures by type and model."

Carl speaks up again. "It sure is a significant difference between the small models and the large ones. Did you notice any other trends?"

"Only a small one on the cracked housings. Here is a chart that will show that." I pass out the chart that Dr. Stromberg suggested I make showing just the spring failures and the housing failures by model size.

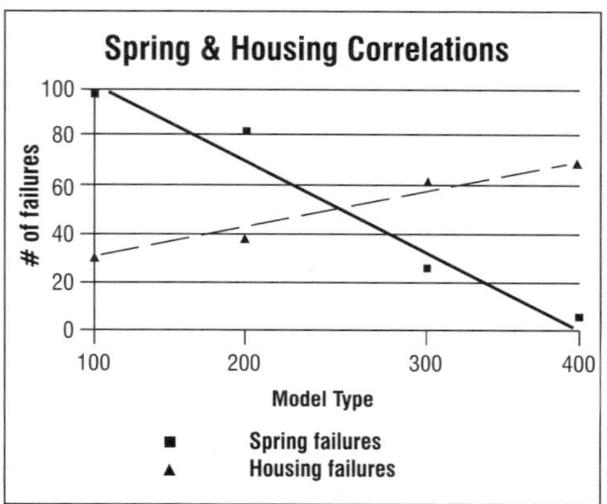

"You can see that the number of cracked housings gets higher as the model size increases so there is a positive correlation there."

Vic pipes up, "Whoa, positive correlation! This guy's good."

Vic and Carl laugh, and I blush a little then continue. "Dr. Stromberg showed me that too. Anyway, other than that there isn't too much that seems to have any pattern to it."

Carl interrupts, "Actually your original data shows there is also some correlation on the model type and number of contact failures. Isn't that true?"

We look at the original data and I see he is right. I admit that I hadn't really seen that as a correlation, but he is right. It could be helpful also.

As everybody looks at the charts I just stay quiet. Dan looks at me and shrugs his shoulders. Finally Vic looks up at me and asks, "This is very good. As a mater of fact it helps justify the design change in the housings we are considering. I think we better take a close look at this data and why this correlation exists as we work on those changes. Maybe Drew can help there when he finishes this project. Anything else Drew?"

"No I don't think so. I really don't know anything about the process or material so I don't know what to do. Sorry."

Amy speaks up now. "That's fine, Drew. Very good job. Dan, do you have anything to add?"

Dan has been leaning his chair on its back legs. Now he drops down to all four. "Not much, only one other thing we talked about is the change to the new material that we now use on the springs."

Carl interrupts, "Good, I was wondering about that."

"Well actually the springs, which as you all know use Springalloy, used to be made of some high grade steel. Now with Springalloy the failures are much less for the large models but did not improve for the smaller models. In fact it's about the same, maybe a little more. So the total number of spring failures dropped, but the smaller models are no better." He stops and leans back again in the chair. "But I'm not an engineer, and neither is Jess. So we can help you with the shop parts of the problem, but other than changing to some other material we can't help solve it."

At this point, Amy, who has been rather quiet speaks up. "The process is still the same isn't it?"

Vic nods his head. "Yep, the same process as before, in fact pretty much the same machines also. We did buy a new one but it does pretty much the same thing, as far as I know. Drew, didn't you go talk to Jess in the spring shop? Did he give you any other information than what Dan had told you?"

"Not really. It's pretty much the same. According to Jess the only difference is that the rod is heated before it is wrapped where before it wasn't. Other than that it's the same."

We hear Carl begin to mumble something to himself. It finally becomes audible, as he says, ". . . the fatigue failure is from the hot wrapping of the wire."

"What's the deal, Carl?" Vic asks.

He realizes we are all looking at him now, and he speaks up. "The springs used to be all cold wrapped. That is how most springs are made and it's the right way. Cold working a material has the advantage of maintaining most of the strength properties we need. When we started using the Springalloy we started hot working the stuff because that is what the vendor called for. Spingalloy is a special material, and we pretty much just follow the instructions of the vendor. As I recall, however, when we had the vendor do the tests on it for us, we only used large S-line models for the tests. I guess we didn't realize that there would be that much difference between the large and small models."

"What's the difference between hot working and cold working?" Dan seems interested in this now as he asks the question.

"Well there is quite a bit to it, but basically cold working is done below the annealing temperature and hot working is above it. The annealing temperature is the temperature at which the material is relieved of the internal energy caused by deforming it."

Amy speaks up now. "I see where you are coming from, but I didn't think our process was really a hot working process."

Carl explains. "Really it isn't. It is a warm working process not intended to anneal but just to help in forming. The temperature for warm working depends on the alloy."

Dan stops them at this point. "Wait up guys. Sorry, but I still don't follow this idea of annealing stuff and internal energy."

Amy seems to want to explain something here. I hope she is not going to stop this discussion, because I am lost just like Dan. I have no idea what these guys are talking about and I'd like to follow it. Finally Amy speaks up.

"Good question Dan, and don't be sorry. You just haven't had a chance to learn this yet. I bet Drew is just as lost as you are." She looks at me and I nod that I am.

"See this rubber band?" She takes a rubber band that was around a set of papers she had. Dan and I both nod our heads. "Now it I put a little bit of tension on it like this." She stretches it. Not real tight but just a little. "I have created some internal energy in the rubber band. Now, other than letting go of the rubber band, how would I lessen the tension?" I'm not sure what she means yet, and I don't think Dan does either.

Carl must see where she is going because he jumps in. He explains. "Other than letting go of the rubber band or cutting it, what else would make the rubber band relax so that it was no longer in tension?"

Suddenly Dan's eyes light up and he blurts out, "Heat it up!"

Amy continues, "Exactly right. If we heat it up it will release that internal energy, what we are calling tension. What happens is that the molecules reorganize and reform to accommodate the new shape required by the force put on the material. Does that make sense?"

Dan smiles widely, "Sure does. That's what happens to me too if I get too much tension and internal energy. I heat up, blow my top, and it's all released. Then I'm fine."

"Ahhh," says Carl, "but what else have we lost if we heat up the rubber band and lose that internal energy?" Dan looks confused but I think I know what he is asking.

"You lose the usefulness of the rubber band." I answer.

They all nod, and Amy asks, "What do you mean?"

"Well,'" I answer, "the rubber band is to hold things by tension. If you release that tension, it is no longer useful. Therefore you want the thing to be able to keep those properties so it stays useful."

Dan frowns as Carl exclaims, "Exactly! And the same thing is true for the springs. The spring is useful due to the internal energy; in other words, its stiffness or ability to recover. If that is lost then we have no spring. For the most part, cold working maintains those kind of characteristics in a metal as long as they are not overdone. So the trick is to keep the temperature at the right point to make it easy to work with, without losing the valuable properties. In the case of the Springalloy springs, the small springs have overheated and lost too much of their tension. When they are worked in the solenoids, they don't have the recovery force to push the solenoid plunger, particularly if it is a little dirty or otherwise reluctant to move. Or the springs just lose resistence."

Dan is still frowning. "You mean blowing off steam makes me weaker? That's not true, is it?"

Amy smiles. "Something to think about isn't it, Dan?"

Dan grunts, "Ahhh, dumb idea." But you can tell he is thinking about it.

Vic speaks up now. "Well, what do we do from here?" He's pretty good about keeping things moving. I suppose that is why he does the job that he does.

Carl answers. "I suggest we do a little experiment to start, and I think I'll call our Spingalloy vendor and explain what we have. They might have some idea on what to do to correct this. Vic, maybe you can help Drew and Dan set up a simple experiment testing the temperatures and spring sizes against elasticity. I don't know what the pre-heat ovens go up to, but we should be able to get them hot enough to learn a little about the pre-heat temperature as it relates to spring size. Dan, didn't you make some device two or three years ago to test springs?"

Dan answers. "Yeah, they're in the back of the cage. I can get them out tomorrow and see what kind of shape they are in."

Then Carl adds, "Amy, why don't you talk to Debbie Anderson in electrical engineering about the burnt contact problem. Though they aren't as bad a problem as the springs, I know she would be interested in the data, and it wouldn't hurt to see if we can reduce those as well."

"Good idea, I'll do that." Amy says, "Can we get everybody back here next Friday to see what we have?"

We all agree we can meet next week and set the time. The meeting breaks up but Vic, Dan and I stick around to talk about how to set up the experiment. We make a decision, or perhaps I should say Vic makes a decision, because Dan and I don't have much to say. For right now, until we learn more, we will just do a simple experiment to see how long it takes a set of each of the spring sizes to fail at three different temperatures. One at the same temperature as they are being produced now, one higher by about 50 degrees and one lower by the same amount.

I'm supposed to go talk to Jess about getting the springs tested at the three temperatures, Dan will check out the test machines he has, and Vic will run it by upper management to make sure they know what we are doing. He will also make sure any material and time used won't affect production scheduling. I wouldn't have thought about that. It's clear from this afternoon's meeting that I still have a lot to learn. It was really neat though, learning about the material and seeing how everybody had ideas and helped come to a decision. I also gained a new respect for Amy. She is really bright and knows her stuff.

Everyone leaves, and I head toward the spring shop. I notice Vic in his office so I stop to ask him a question. "Hey Vic. Tell me about Carl. He really seems to know his stuff."

Vic smiles. "Carl is an interesting guy. He has a BS degree in physics, then he got a masters degree in manufacturing engineering with an emphasis in materials. He's the best materials man we have here. He came about four or five years ago and has really helped us a lot. Why do you ask?"

I shrug. "I don't know. He's just an energetic guy and seems to understand things well. He seems pretty bright."

"That he is," Vic answers, "bright and hard working. A good man for problem solving. Between his understanding of physics and his engineering training, he thinks very well."

Vic needs to leave so I head toward the spring shop to find out from Jess how we can get the springs for our test. I really felt dumb as he asked me how many we would need of each size, when we needed them, and what machine we would want them from, or did it really matter. I realized that there were a lot of questions I hadn't even thought about. Jess said he could do the runs probably during the lunch hour on Monday so as not to affect production.

I run back over and talk to Vic about the details. I want to get back to Jess before he leaves for the day so we can get the springs by the middle of next week and get these tests done. Finally I have the information I need and hustle back over just in time to get Jess. He says he thinks he can get them for us by Tuesday morning. Just as I am about to leave, Jess asks, "Drew, do you have a minute?"

I tell him I do because I don't usually leave till 5:00, but I know his shift is over at 4:00. He comes in at 7:00 in the morning. He assures me he is fine and asks if I'd like to sit down. I accept and we sit at his desk.

"Drew, you seem to like this stuff. Are you planning to get a college degree in engineering?" He's looking at me like my dad would.

"Yeah, I like it. It has been fun, but frustrating. I know so little I feel like I can't help much. I'm not sure if this is what I want to do either. I mean, will I make enough money? Am I smart enough to make it? Will I get bored? Man, there are just a lot of questions. I don't know."

He laughs at that then leans back in his chair. "Drew, there will be a lot of questions and that's OK for awhile. As for being smart enough, that's not really a question. Sure you are smart enough, most everybody is. It's not how smart you are but how hard you are willing to work. There are a lot of smart people working out in the assembly area and shipping that just didn't want to put in the effort. As far as making enough money? Engineers make good money. You probably won't get super rich doing it, but you'll do well and will have what you need. It's not a bad life. Good benefits, a lot of respect from others because of the kind of work you do and a good challenge. I've watched the engineers here. They do well. There are a couple who aren't very well liked in the shop but that's because they don't treat others with respect. You won't be like that, I can tell already. By and large, all are well liked and do good work. You ought to give it serious consideration Drew. You would do well."

Maybe I'm being a little too bold but I can't help but ask, "Jess, I don't mean any harm, and I appreciate your encouraging me, but if you feel that strongly about it, why aren't you an engineer?" I hope I didn't hurt his feelings, but he doesn't seem hurt. "Besides, the work you do is important, isn't it? And you seem to enjoy it, and it appears you're very good at it."

He looks at the ceiling for a minute then back at me. "That's a fair question Drew. Listen, if I had put the effort into it I probably could have been an engineer. I won't go into detail, but things just didn't work out for me. Partly my fault and partly the way things were with my family. Education now is critical. These days it seems like you need to have a degree to really have a good chance of making a mark in the world and progressing in whatever company you choose. Sure a person can make it without the degree, but the degree increases your chances."

He now leans back on his chair legs like Dan did in our meeting. "As for my work, yeah I do enjoy it and I am pretty good at what I do. But it's kinda like being in the army, Drew. You need a good sergeant to keep the troops in line and run the show. You can't do it without them. You also need a good lieutenant. That's the guy who has been trained in a broader view of the role of the troops and how to carry on the battle. I'm a great sergeant, Drew. You could be a good lieutenant. How you worked with and listened to Dan in solving this spring problem is a good indication that you would do a great job."

He sits his chair back down on four legs and adds, "The same is true anywhere for that matter, Drew. If you put in the work, you can be the guy who has the broader view and can lead the sergeants and the troops in the battle."

He stands up now and leans against the desk. "Well, enough preaching for now. Hope I didn't bore you to death. Better go. Don't want to get stuck in bad traffic." He gets his lunch box as I stand up to leave.

"Hey Jess," I call. He turns around. "Thanks a lot. I appreciate you taking time to share some advice."

He shrugs his shoulders, "Well you know about advice don't you?" He smiles then continues, "It's free, and worth at least what you pay for it." As he turns and heads out the door he calls back, "Have a great weekend and I'll get those springs for you by Tuesday."

I think about what he said all the way back to my desk. His advice is certainly worth a lot. I guess I do believe I can do this, the quesion is do I want to? Is this the thing for me? Maybe I worry too much.

When I get back to my desk, Vic is in his office going over the new prints. As I pass he calls out to me. "Hey, what did Jess say?"

I turn back and stand in his doorway. "He said he could have them by Tuesday morning. I hope he does."

Vic smiles and says, "Listen, if Jess said Tuesday morning they will be ready Tuesday morning. When Jess tells you something you can take that to the bank. They don't come much better than him."

We spend a few minutes going over how to conduct the tests and then Vic wants to leave a few minutes early. His son is playing in a football game tonight and he wants to go home and get a bite to eat before they go to the game.

I take a little time jotting down some more notes in my work journal, then wrap up and head home. What a day to end the week on. Looks like we are on the way to getting the problem solved and I got some great advice. Not only that, but according to Vic, what Jess tells you, you can take to the bank. Maybe I'll take Jess's advice and bank on it. I can't think of a better thing to study right now.

Well, it's the weekend. Tomorrow a bunch of us will go to the football game then relax and get ready for next week. I have three tests next week. If I'm going to be a good leader of the troops, I better make sure I learn the kind of habits I need to have. Good people at ACTC. I'm sure lucky to have this job.

Chapter 15

The weekend was OK but a little hectic. I wanted to get some studying done for the tests I have coming up this next week. I wanted to take time to relax some, too. I did get to go hiking in the hills near campus. That was a good break. The weather is starting to get a little cool and I'm looking forward to the snow. One of my goals is to learn to ski, and as soon as the local ski resort opens, I can get started on that; that is, if I have time and money for it. It is clear that if I want to, I can put in more time at work. From the standpoint of picking up a little more money that would be good, especially since I have to come up with the extra for housing next semester. It would also be nice to have the money to take the ski lessons, but then working more means less time for other things. I think this is called real life—how to balance it all and where to put the time and money.

Troy, Russell, and I also spent some time studying together for the physics test. Russell has a pretty good understanding of the material, and I think Troy and I probably got more out of it than he did. He seemed grateful though. He said reviewing it helps him better understand and remember it. The same is true for me. Troy and I talked about Kellie and Christy also. He has been out with Kellie a couple of times and he said things are going well. Not exclusively dating each other, but they enjoy being together. Troy suggested we get together again for a double date. We had a good time the last time we did that, and it's always fun to get a group together. More people to talk to and get to know.

This time the physics test was offered in the testing center so we could take it anytime the first three days of the week. I felt ready on Monday, so I went ahead and took it. The testing center was almost completely empty. Even the desk attendant looked surprised when I asked for the physics test. I guess most everybody is doing what I did on my last physics test. I learned my lesson, at least this time. I got a 91.5% on it. That's pretty good but, as usual after I got it, I wanted a better score. Never satisfied I guess. Anyway, it was a lot better than the last one.

In Dr. Stromberg's class we learned about what he called engineering economics. It made a little sense, but since I felt confused I decided to go back to my room after class and review the notes I had taken.

I really have been trying to follow the stuff I learned in the class at the beginning of the semester on good study habits. It has helped, but I find the hardest thing is to stick with it consistently over time. I'll do well for a few days, then contrary to what I was told in the class, I reward myself with slacking off. That's not the way it is supposed to be done. I suppose that is why they call it developing discipline and character. I was complaining to Christy about how easy it was to make a goal and how hard it has been to follow through with it. She printed up a little quote for me that said, *["Character is the ability to follow through with a decision after the emotion of making that decision is past."]* I put in on my mirror but I'm still not sure I like it. It makes me remember my goals too often. But I guess that is a good thing.

Try as I might I can't quite get into this stuff. Suddenly the door opens and in rushes Stu. "Hey Drew, what are you doing?" We are both surprised to see each other because I usually study in the library, not in my room.

"Oh, just reviewing some stuff from class today. I can't quite get it through my head. What are you doing? Do you usually come back to the room at this time of the day?"

He throws his backpack on the bed and drops down after it. "Yep, usually. This is when I get my homework done. Hah, you didn't know I did homework, did you? What are you working on, more physics stuff?"

I shake my head. "Nope, it's engineering stuff, at least that's what the professor said. It's engineering economics, stuff like rate of return, risk, and cash flow. Seems like it shouldn't be too hard to get but I'm just not concentrating, I guess."

"Whoa, what did you say?" He sits up straight now.

"I said I guess I'm just not concentrating. Why?"

"No, before that. What are you working on? Engineering economics stuff?" He stands up and walks over to my desk now and looks at my notes. "That's what I thought you said. Hey, you helped me with physics, how about I pay you back and help you with your econ? Deal?"

I'm not sure how he can help, but he seems to want to. "Sure, whatever."

"OK, what don't you get?" He has pulled his chair over to mine now and seems pretty excited about this. "How about risk and return? Do you understand that?"

I think I do but I could use the brush up so I tell him, "Go for it."

"OK. First, return is simply the amount of payback you get from how you choose to use your money. Let me give you some examples."

I interrupt him. "What about risk? I think that's part of my block right now."

"Just wait. I'll tell you about risk after these examples. OK, if you have your own personal money, let's say a thousand dollars, and instead of spending it you want to get some kind of gain from the funds, you have several ways you could possibly do it."

I interrupt again. "You mean like give it to you?" He looks frustrated. "No smart aleck, you wouldn't get anything back from that. Are you going to listen or not?"

"OK, sorry." Man, he is into this. Maybe he knows what he is talking about. I thought he just goofed around all the time. Anyway, he continues.

"First let's say you just decide to put it in a pop can and don't touch it. It is very liquid, but the return will be very low.

I can't resist. "Is it liquid because it's in the pop can?" I laugh but he doesn't.

"Listen, son," I think he is joking now, too. His voice deepens and becomes very formal, "You want to learn this or not?"

"OK, professor, yes I want to learn it. Liquid means easy to get to. Right?"

"That's right. Money put into a pop can is very liquid because you can just walk over and get it. If you bought stocks with it, it isn't as liquid because you would have to sell the stocks first to get at it. Anyway, you don't get any return if it is in the pop can because it isn't earning you any money back. In fact, you lose money due to inflation. So that's example one." He writes on my notebook page:

1) $1000.00 in the pop can — very liquid — no return — very safe (low risk)

"Wait" I tell him, "what do you mean 'low risk?" He holds up his hand as if to stop me. "I'll tell you later, be patient."

He starts writing the second example. "Next, you put it in a savings account. It is still pretty liquid, you just have to go to the bank to get it. The return is fairly low, but better than a pop can. It will probably keep up or even slightly exceed inflation and it is still pretty safe. That means unless the bank or economy fails, you will get it back." So the second one in my notebook looks like:

2) $1000.00 in the bank — still liquid — low return — very safe (low risk)

"OK now for the third example let's say you put it in a certificate of deposit, called a CD. In other words, you're loaning money to the government for a set time period. This still is similar to the savings account, except in terms of liquidity, because if you take it out there is a fairly substantial penalty. But the return is better, probably half again over the savings account. So the third example is . . ."He writes:

3) $1000.00 — less liquid — medium return — still safe (low risk)

"OK. For the last two examples you invest your thousand in stock. In both cases it is less liquid because, like I said earlier, you have to sell the stock to get it. The return is almost always quite a bit better. The difference between the fourth and fifth examples is that in the fourth one you buy blue chip stocks. Blue chips are established companies that are still fairly safe. But they are less safe than a savings account, because if the company loses money, so do you. If it fails, your money is gone. The return, given that the company does OK, will be good, certainly higher than a savings account, and will likely be at least twice what your CD earns.

Example five is buying what are called speculative stocks. These high risk stocks promise very high returns if the company hits pay dirt. It's like an investment in a secret gold mine. If the gold is really there you will be rich, if it is not you will lose every cent." Examples four and five look like:

4) $1000.00 — in blue chips stocks — not very liquid — good return — fairly safe (low to medium risk)
5) $1000.00 — in high risk stocks — not very liquid — very high return — unsafe (very high risk)

It makes some sense like this. "The risk is basically just how likely the investment is to fail, is that right?"

He nods. "Basically that is right. So you see that the higher the return, usually the higher the risk. So what you want to do is to try to maximize your return and minimize your risk."

He leans back in his chair. "What you must ask in all cases is, 'Is the extra return worth the risk?' Well Drew, my friend, does that make sense?"

I lean back too. "Yeah, I think it does. I'm still not sure why I need to know this. But my professor says it is to understand and do return on investment calculations. He just mentioned that today but we didn't do much."

He sits back up. It is obvious he likes this stuff a lot. "ROI — that's what return on investment is commonly called— is just figuring out how much money you can make on an investment and how long it will take to have the money you invest in something, like a new machine or something, pay itself back." Now he has written ROI and its definition in my notebook.

"As far as why you need it . . . in your work as an engineer you will need to make decisions about equipment you might want to buy, or justifications on building a new plant or starting a new product and stuff. So knowing this can help you do that."

"Yeah, Dr. Stromberg did say that making accountants happy was important. They approve the money, or at least go over the requests for it, so knowing how to give them reasons why it ought to be spent is important."

Stu leans back in his chair again. "Even something like spending money for training employees on something new needs to be considered from a financial viewpoint. This is one we talked about just yesterday in my financial management class. You will need to determine the costs, when and how the funds will be spent, and then calculate the amount of return for the money spent and how long it will take to get the return. Do that well and you'll have a lot better chance to get your money. By the way, most companies have a set limit that a project will need to return if the investment is to be considered for approval. Most require a return substantially better than the prime rate. Prime rate is a measure of how much return a bank gets. Also the payback period should be around three years according to my business professor, but the company and the type of project it is will make it vary."

He gets up and goes over and lays down on his bed. "Wow, that was tiring. I'd better take a nap. Does that help, Drew?"

I need to get ready to go to work so I start putting my stuff away while I answer. "Sure does, Stu. Thanks for taking time. You know this stuff pretty well.

Are your classes going well?" I really am impressed with how well he knows this stuff.

He's rolling over now. I thought he was joking about taking a nap but it appears he is serious. "Classes are fine. We can talk cash flow later if you want. Fun thing to think about though. It was a test question last week in my financial management class – if a company is making a profit through sales, how could it still go bankrupt? See you tonight." With that he closes his eyes and is out in no time. I've never seen a guy who can konk out so fast.

Before I finish putting it all away, I take a look at the first page of my notes from class about engineering economics. The notes make more sense after thinking about things and going over this with Stu.

Introduction to Engineering - Lecture 13 - Engineering Economics
Money

— critical resource
— a common constraint in eng. & tech work

Knowledge of economics will help:

— understand financial statements generated by your company
— manage budgets for projects or groups you will supervise
— use resources entrusted to you better
— communicate better with financial departments and people
— help you make better decisions regarding project or product payback and worthiness
— prepare you for management

Key points

— Return
— Risk
— Cash flow
— Leverage
— Return on investment
— Payback period
— Time value of money

I finish putting away my stuff and still have twenty minutes before I need to leave for work. Since I have a test in my art history class I pull out that text and review a few things I underlined from the last class before I take off. Interesting, I didn't know that Leonardo da Vinci designed the first helicopter.

True to his word, just like Jess said, when I get to work there is a note on my desk.

> Drew:
> The springs you wanted are in the test shop with Dan. He wanted to wait until you or Vic talked to him before he started testing them. Said he didn't want to blow it. I have the records of the runs like you said you guys needed. Stop by when you want to get them.
>
> Good luck, Jess.

Just then Amy comes in. She sees me and says, "I understand Jess made the springs. Before you start the tests I'd like to talk to you. Do you have a minute?"

"Sure, I just got here and was going to go see Dan. What can I do for you?'

She motions for me to come in her office and sit down. "By the way, did you know it took Jess almost three hours overtime last night to get the springs done?"

"No, I didn't. I didn't talk to him yesterday because he said he wouldn't have them until this morning. What happened?"

"I guess one of the spring machines broke yesterday, so he wasn't able to run our test springs until after he got the regular production ones done. He had figured to be able to run them during the day on the new machine after the regular run. But he needed the new machine to finish the production run, so he had to stay after to get ours done."

I feel a little bad. I hope I didn't make him feel like we would be mad. I tell Amy that.

"Oh, it's not your fault. It's not anybody's fault. That is just the way Jess is. When he says he will do something, he does all he can to make sure he follows through. A great guy that Jess." She shakes her head a little. "If only everybody was more like that, we would have a much easier time with each other." Then she looks up at me. "Dependability and honesty. Drew, no matter how smart you are, if you aren't dependable and honest you won't have the respect of good people. Remember that."

Boy, that's the same thing Vic said about Jess, and I've heard Dan talk highly of Jess too. I think I just got another lesson in the characteristics of being a successful person.

I don't think Amy goes to the spring fabrication shop that often. I wonder how she knows all of this. I am still a little apprehensive around her just because I don't know her well yet. I ask, "How did you find this out? Did Jess tell you?"

She shakes her head. "No, I doubt Jess would have mentioned it. Dan came to see me, and he told me."

"Dan came to see you? Wow, he told me he doesn't like to come up here. He says he's afraid someone will slap a tie on him and sit him down at a desk. Says he'd rather have his head in a vice. What did he want?" Evidently my description of Dan's comments entertain Amy. I think she almost laughs. She does smile pretty broadly.

"Well, I guess my comment last Friday in the meeting comparing his letting off steam to a material losing strength made him think. He came to ask me about it. We had a good talk. I understand him better now and maybe he learned something, too."

What she says makes me real curious, but I know enough not to ask any more questions. Whatever the case, I suppose I will see how it came across to Dan when I go to the test shop. She is opening her desk now and pulling out a couple of sheets of paper. She puts one of them in front of me.

"Drew, before you start the test, you need to make sure you have designed it correctly. I wrote down a few notes and had Vic go over it. He has a copy and this copy is for you. Follow these guidelines when you do the tests to make sure the results are accurate. By the way, do you keep a design log yet?"

I think she is referring to the notebook with notes about everything that is going on. Vic told me about it. I hold it up. "Yes, Vic suggested a couple of weeks ago that I start keeping this."

"Good. Keep good notes. It might seem like a waste sometimes, but it can be very helpful. This other sheet," she passes it over now, "is for Dan. Give it to him so he will know what we have told you and what the important characteristics of a good experiment are. He probably is familiar with most of this because he has heard us talk about it before. Let's take a look at it so I can explain some of these things."

Amy points to the first heading about documentation. "I think we have said enough about this. The next thing is the most important in terms of getting good results. For example, I know that Vic told you to have Jess run all the springs for this experiment on the same machine. Do you know why?"

I wasn't sure then, and I'm still not sure so I tell her that. I add, "It seems to me that we might want to have a variety of springs from different machines so we could cover all the bases. So I guess I'm not clear on this one."

"Good. Let me explain. If we were trying to test the machines or to see the differences of uniformity of springs from machine to machine, then we would want to have springs from all the machines. But the focus of our test, as you see from this point," she points to the last one under item 2, "is to test the material or process of the small springs, not the machines. Does that makes sense?"

It does now. "Yes, it does. Since we are testing the springs and not the machines, then we want them all from the same machine to eliminate any variation that could be caused by one machine versus another. Right?" I look at her as I finish what is as much a question as a statement.

"Exactly. We want to vary certain conditions in making the springs so that we will understand those variables better, but we want to exclude all other variations. Therefore a critical part of this experimental procedure is to eliminate the different kinds of variation you might see. The special variation, or gross error, that is listed

1 Document the experiment as well as possible. It helps validate the results and makes it much easier to remember later on and recreate it if necessary. Record things such as:
- times
- locations
- machines used
- operators
- temperatures
- environmental conditions
- your own observations

2 Reduce sources of variation as much as possible. For example:
- use the same machines if at all possible (if you can't, see documentation above)
- perform experiments or tests under the same conditions
- follow the same procedures for all tests and experiments
- eliminate gross or special errors (errors from sources external to the experiment)
- reduce system or common errors (errors that are a normal part of the system)
- keep focused on the purpose of the test by reducing outside parameters

3 Use necessary time to complete the experiment well and fully.
- don't rush any part of the experiment
- don't leave out planned parts of the experiment
- changes should be thought out and discussed with others involved

4 Report the results of the experiment.
- written report
- list assumptions clearly
- submit to document control
- distribute to all involved in the experiment or affected parties

here is variation caused by things other than things inherent to the process or machine we are testing. Those kinds of things we can usually control or eliminate if we are thinking about them. For example, if the testing device has a problem that is due to a faulty controller, then that is a problem that will cause faulty results but has nothing to do with what we are testing. In fact, it is an error introduced by factors external to our tests.

"The other kind of variation that occurs is systematic or common cause variation. It can be very difficult to control and even identify. For instance, perhaps there are minor differences in the alloy or the diameter of the wire, or a small variance in the process that made the spring. Even, to a certain degree, the normal variation of the testing machine itself. These usually are evened out by repeating the tests over numerous pieces. This is why we are going to test 40 springs for each condition, not just one."

She looks up at me and seems concerned that I know this. "Does this make sense?"

Frankly I'm not sure but I give it a try. "Well, I think so. If we have differences in the results because the testing machine was bumped or knocked off the table, that's this gross or special error. On the other hand, random errors just happen. Is that right?"

"Sounds like it. Good. It appears you are catching on. Now, the last two items are again housekeeping kinds of things that are directed at keeping the experiment clean and organized. Also, the reporting item lists what is expected at the end of the experiment. You might not be that excited about writing a report, but it is an important part."

I nod, knowing that it is. I have also wondered about using this as a chance to do the report I need for my intro to engineering class. I decide this will be a good chance to find out. "The report is fine. I'm just not sure that I can write what ACTC really wants. I will want help making sure I do it right." Amy seems agreeable.

I continue. "Ah, I'm supposed to write a report for introduction to engineering and engineering technology class. Do you think I could use this project for my report for school?"

Amy appears thoughtful for a moment, then responds. "I am sure we could work that out. We just need to make sure, if proprietary information is divulged in the report, that whoever reads it agrees to keep it confidential. Companies are always very concerned that any company information does not get to competitors. ACTC is no different in that regard. There are a lot of other companies who make products like ours, and we want to keep any edge we have to ourselves. I am sure we can work that out. Any questions?"

I remember Christy's interest in talking with her. "Yes, just one. I have a friend at school, her name is Christy, who is interested in learning about what you do and how you feel about it. Would it be OK to have her come and talk with you?"

Amy smiles. "Of course. Just let me know beforehand when she will be coming, and I would be happy to visit with her. She could talk with Debbie Anderson in electrical systems and maybe even Julie Harper. Julie manages a group in technical sales. Between the three of us we can probably answer her questions."

She asks if I have other questions. I tell her I don't, but thank her and then excuse myself from her office. I'm excited to tell Christy. She's a neat girl and it would be great to have her in some of my classes. Between her, Troy, Russell and myself, we could get a pretty good study group together.

Before I go to the test shop I want to review the page she gave me so I know what I am giving to Dan. I still don't feel confident that I could answer any questions. But like Amy said, he probably knows most of it already. She clearly knows what she is talking about and also explains it well. The more I get to know her the more I like her.

When I get to the test shop I can see nine boxes on the counter. They are labeled Drew's Springs, and each one has the temperature at which the spring was made and the model size for the spring. With 40 springs per box, I wonder how long it will take to test them all. I have no idea how Dan proposes to do this, but he said he had a machine to do it. Dan, who wasn't here when I got here, walks in and sees me. "Hey college boy, whatcha up to today?" If his talk with Amy calmed his exuberance it's not showing now.

"Doing well, Dan, how about you?"

"Dandy, my friend. You ready to start these tests?" Without waiting for me to answer he continues, "Listen, I got the test machines out yesterday and cleaned them off. This morning I tested them out and one of them works fine but the other is acting up. I think it is an electrical problem so I have Chris Johnson working on it. He is a great electrical technician who can fix any electrical problem you can come up with. In fact he went to SU, also. He studied electronics technology or something like that. You ought to talk with him sometime."

Thinking back on what I just learned from Amy, I wonder if we ought to use both machines or not. I remember that I have the sheet from Amy for Dan so I pull it out of my notebook. "Dan, this is from Amy. She went over it with me and said to give you a copy. It's to help make sure we conduct the experiments right."

"I saw her this morning and she told me she was getting it together." He pauses as he looks it over then asks, "Do you know what all of this means? Some of this makes no sense to me."

"Well, she did go over it with me. I think for what we need to do, we just need to make sure we don't do anything that may add variation into the test."

"Whatdaya mean?" he asks. "What variation are we adding?"

I hope I say this right. "Well, for example, in the second part about variation, I think it might be best if we only use one testing machine rather than two. That way any differences between them won't affect our tests."

He doesn't look mad. "Oh, I see. You mean if they operate at different forces, that could affect the test in a way we don't want it to. Yeah, that makes sense. But it will take longer to finish the tests. What's more important?"

He has a good point but I think it is answered as well. "The third point is to make sure that the test is done fully and completely and to not rush it. I think it would be more important to use the same machine."

"Fine with me. But won't we need both machines in the future to test other springs if changes are made?"

He has a point, so we agree to have Chris go ahead and see if he can fix the second machine even though we won't use it now. Also, we conclude that if one breaks down, we'll have the other. He seems relieved to not have to use two machines. "But," he adds now, "it'll take longer."

"How long?" I want to have the results by Friday's meeting if possible.

"Well . . ." He goes over to the desk and sits down with a calculator. "I can test 40 at a time, and it takes about 1 second per cycle per spring, Amy said to run them through 7500 cycles, and we have 9 batches. So that means, . . Let's see, 7500 seconds, divided by 60 means 125 minutes, divided by 60—means a little over two hours per batch—so 18 hours not counting set up and take down. Figure three hours per batch."

I do some quick figuring and so does Dan. That means one batch today. Maybe not, because he leaves in two and a half hours. Anyway, before I'm done figuring Dan speaks up. "Sorry Drew, I don't think I can have them done till Thursday afternoon. That doesn't give you anytime to figure up the numbers before you come in on Friday. Now if we use both machines, I could be done tomorrow. What do you say? It's your call."

He is looking at me now. I really would like to have these done by Friday, but my gut feeling is that I had better stick with one machine. I wonder if Vic is in. Maybe I'll call and see what he thinks. I get on Dan's phone and call but Vic doesn't answer. Amy does answer but I am almost afraid to ask her. I go ahead and explain the situation to her. She asks me what I think. Though I am really nervous about what to say, I tell her I think it would be best to use just one machine and take the extra time. She agrees and tells me that she will talk to the others about rescheduling the meeting.

"We will just use one test machine," I tell Dan. "Can you start tomorrow?"

"You got it, college boy, but why tomorrow? Why not now?"

"Well, you said about three hours per batch, and you leave in just a couple of hours." I feel bad about Jess working overtime. I don't want Dan to do the same.

"Hey, come here let me show you something." He leads me around the corner to another room. He stops and holds out his arm like the girls on a game show showing off a big prize. "Ta da! The machine."

It's a neat looking thing. About three or four feet long, only about a foot wide. It has arms, like little teeter-totters, sticking out from the middle to each side. Each arm has a ringed end that is over a slender rod that I guess is what the spring loads onto. While I'm looking at it he gets my attention and asks, "Well, what do you think?"

I guess I had been staring at it. "Hey this is cool. You made this?" I think I sound a little too surprised.

"You bet I did. Awhile back we wanted to do some tests on new springs and so I built this little sweetheart. I got the idea from looking at the rocker arms on a Chevy small block cylinder head. See, these arms lever up and down two times a second and with a spring on each side that equals one cycle per second per spring.

Now, about my quitting time that you are worried about. Up here," he points to the control panel, "I can program how many cycles it needs and off it goes. When it gets to the end of the count, it shuts off automatically."

I am really impressed. This looks like professional work, not some shop made thing. I realize that this *is* professional work. "This is really impressive, Dan. You did a great job. How do you know when each spring fails?"

"Good question. Each arm has its own sensor and counter. When a spring fails, or at least gets below a certain resistance, the counter stops and is recorded in the computer. Then I just read each one and record it. So, my friend, I will have the first set done tonight, and I will take the readings in the morning."

This is pretty good stuff. Dan and I go over the rest of the stuff that we need to keep track of. For example, we decide that if the tests are going to run past his quitting time, the room temperature should be kept the same. He will talk to maintenance and make sure the thermostats are set the same all night. We finally feel that we have it all set, but I'm going to double check with Vic when I get back.

He starts to get the machines ready and the first batch set up. I watch him for awhile then say goodbye and turn to leave. As I leave he stops me, "Hey, college boy."

I stop at the door. "Yeah, Dan?"

He fidgets a little then just says, "Thanks, Drew."

"Thanks for what?" I'm not sure what he means. And he called me Drew. . . seriously.

He pauses for a minute. "Thanks for including me in this. It's been fun and I've learned a lot. Just thanks."

I shrug my shoulders, "You're welcome, I guess. I mean, you've done as much as I have in this. Thank you, too."

He gives me the thumbs up and I turn and leave.

Dr. Stromberg said something in class the other day that comes to mind as I leave. He said sometimes people work so long in a place they feel that they get taken for granted. They don't feel challenged or appreciated. This particularly happens to the hourly people who haven't had as much education. Maybe that's what has happened to Dan. Maybe to Jess too, but it seems he has a pretty good handle on life. Whatever the case, I am learning that everybody has a role to play and all the roles are important. In fact, as I think about it, getting to know these people at ACTC and working with them has been as good as the technical experience. Dr. Stromberg also said that we should make sure we know that people care about their work. He said, "People won't care how much you know until they know how much you care." Maybe he's right.

Chapter 16

The rest of the week passed without much incident. Amy changed the meeting time so we will be meeting later today, Monday. Dan was able to get the tests done by Thursday afternoon. He actually did have to work overtime one day because of a malfunction in the test machine, but his supervisor approved it and the tests were finished in time for me to put all the data together and make more charts showing the results.

I spent much of last week learning more about the CAD system. With the labs at school on CAD and the time spent here, I'm starting to get the hang of it. I'm not sure when I'll be using it for real work, but Vic is determined that I learn how to use it well. It has gotten a lot more interesting since I have learned what it can do. It has the capability to organize and relate all kinds of drawings and information as well as make actual three-dimensional models of the part for different kinds of analysis. The analysis can be performed even before the part is built, so changes can be made that hopefully will make a better part when the actual physical model is built. I'm just getting to the modeling part. I haven't gotten into the analysis stuff at all, but I can see how this will be a great help. Vic says the problem is that too often tools like the CAD system are used below their capability. Well, that's all been fun, but now back to the spring problem. It's Monday and hopefully it's the day we get to the bottom of the whole thing.

I spent all Friday compiling results of the test. They look pretty interesting, but in some respects there are no surprises. Some parts of the tests show what looks to me to be the way we ought to make the springs. I'm learning though that there is often information that comes out in our meetings to change that impression.

As I am sitting in the conference room waiting, Stan Hall walks by. I wave when I see him go by, and he stops and comes back. "Hello Drew, how are you today?" Without waiting for me to answer, he continues. "I hear you are really getting into this S-line spring problem. Are we near a solution?"

"I hope so. Jess from the spring shop got the test batches of the springs done last week, and Dan got the testing done on the springs last Thursday. I've got the results plotted, and today we meet to figure out where we ought to go from here. This has been a lot of fun."

Stan smiles as he leans against the door frame, and says, "Good, I'm glad you have enjoyed it. What I have heard from Amy and Vic is that the work has been good and it may be a real benefit to the company. We sure appreciate your work."

Just then Vic walks up behind Stan. "You going to join us today?" He slips by Stan. "You're welcome to sit down and help us out."

Stan turns his attention to Vic. "No, I go to enough meetings. Besides, I'd probably lower the level of thinking of this prestigious group. I wouldn't want to slow you down. I'm sure it will be a great meeting."

"Well," Vic's smiling now too, "it will be a pretty high level meeting, and you know what they say—it takes a pretty good meeting to be no meeting at all."

Stan straightens up and throws back. "That's true Vic, but remember, if you stop meeting, soon there will be nothing to meet about." With that he wishes us well and walks away.

I look at Vic a little puzzled, "Isn't it good to have nothing to meet about?" I ask.

Vic laughs, "Not if there is something you should meet about. You see, Stan runs pretty good meetings, and he meets regularly about staff items and such. His point is, if you stop meeting, the things you don't meet about, that you should meet about, won't get done. Sometimes, as frustrating as meetings are, they keep important items on the front burner. It's too bad Stan can't stay. He has terrific technical insight."

"Really?" I say, "I didn't think that an upper level company manager would be that good in technical things."

Vic smiled and said, "Stan was trained in engineering management. That means he has a broad technical background plus a good grounding in business. He uses his technical side and his business side to really serve our customers well. Stan has put it all together."

I change the subject. "Vic, this group seems to work together pretty well. Is it always this way?"

Vic laughs, "I'll say not. But you're right, this group is working out pretty well."

I'm curious now, and ask, "Why is it working out so well, then?"

"Well Drew, I think there are a few reasons. First, the make-up is right. Amy has good representation from all the critical areas—me from manufacturing, Carl from engineering design, Jess and Dan from production shops, and then Amy herself. She represents management and communicates to other areas like accounting and marketing, as well as with upper management. But another reason is that everybody does what they are supposed to outside of meetings. Good groups work well when the individuals take care of their work outside the group. Most of the real work, even for groups, is taken care of outside of group meeting time. That way, we can use our time together more productively."

Just then Amy and Carl walk in togther. Right after them, Dan walks in with Jess. I am glad they are both here. They have been a big part of the whole thing. I have only been here a couple of months or so, but these people have become pretty good friends to me, even if they are quite a bit older than me.

Amy is again the one to start the meeting. "It's good that we can all be here. Thanks for coming. Having representation from all the areas should help us come to some resolution sooner." She looks at me now. "Well Drew, you have the results of the test, I think. Why don't you share those with everyone first? Carl says he has talked to the Springalloy vendor. His information will be helpful. Jess and Dan, any ideas or insight you can add along the way will be helpful." She motions toward me, and says, "Go ahead, Drew."

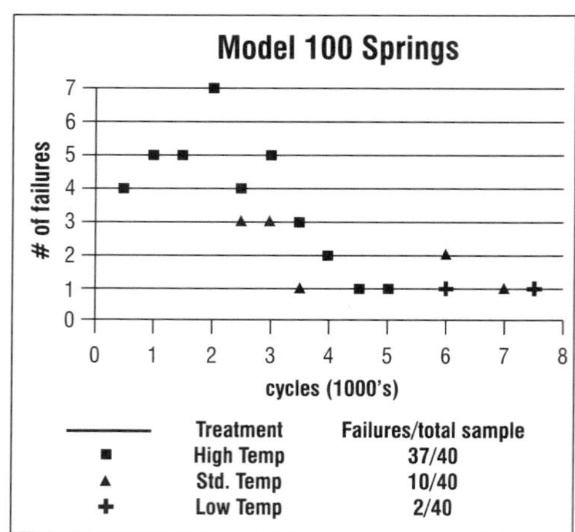

"OK, Vic suggested I make overhead transparencies of the results so we could all look at them togther." I move to the front of the conference room. This makes me a little nervous, but I know I don't really have to say too much, just show the results. I put up the first chart that shows the test results of the model 100 springs at the three different temperatures.

I explain, "I was going to do a histogram chart, but Dr. Stromberg, my professor at school, told me to try this scatter plot kind of chart. I hope it makes sense. Here," I point to the legend on the bottom, "you can see the squares represent failed springs that were made at high temperature, the triangle symbols are the ones made at normal temperature, and the plus symbols are the ones made at low temperature. The chart shows how many springs of the 40 tested for each temperature range failed. As you see, all but 3 of the high temperature springs failed, but Dan said even those that didn't fail were close to it and had pretty much lost their strength."

Dan chips in here. "Not only that, but those failed by just giving out. They didn't break, but they didn't have any ummph left in them. Their strength just got to the point that they failed due to compression values less than spec. Some of the others actually broke but most of them just gave out. In fact, now that I think about it, of the 10 standard temp springs, I think 7 of them just gave out and 3 snapped. Both of the low temperature springs failures snapped."

I nod to thank Dan. "Like Dan said, the failures on the normal and low temps were different and less frequent. A total of ten in the standard temperature range, and only two at the low temperature."

They all look at the chart for a while. I'm not sure if I should say anything more right now or not. Vic finally speaks up. "Is 25 percent failure what we have right now on the model 100 springs? That seems much higher than what we have currently."

Dan sits forward to answer. "No, they're not that high. Actually closer to 12 to 15 percent. But I know from talking to the customers that we don't see all the failures. Some customers just buy a new part. And remember, these were tested to 7500 cycles. A lot of our parts don't hit that before they are replaced."

Vic nods and Amy asks, "Were these all tested on the same machine?"

Dan and I look at each other as I answer that they were. She nods her approval.

Finally Carl speaks up. "Drew and Dan, the way you have organized this is very good. Look at the correlation of temperature to number of failures, and based on what Dan and Drew have said, there is also a correlation to the type of failure. Fracture failures at the lower temps, and just 'no uummph failures' at the high temperature." He chuckles at Dan when he says this. "If the spring has no strength and," he points to the data on the screen, "it appears that is what happened to the ones that were produced at high temperatures, then it appears to me that the material has been annealed, thus the loss of strength. It seems to me the answer, based on what you see here, is to make them all at the low temperature. Vic, you're familiar with processes also, what do you say?"

Vic speaks cautiously. "Well, it appears that way, but . . . ." He stops and looks at me, then continues speaking. "Why don't you show us the rest of the test data first?"

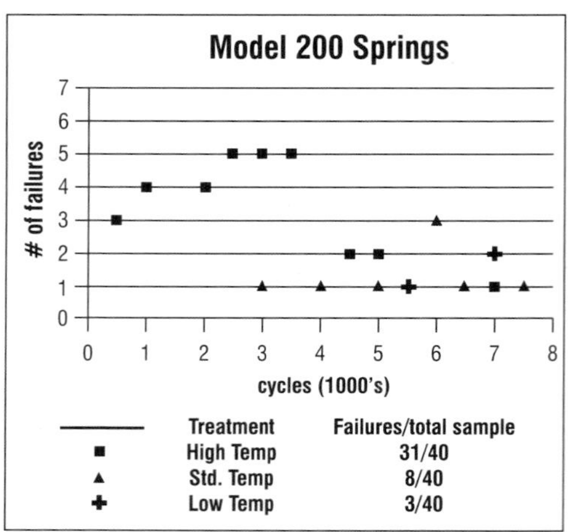

I nod and take the first overhead off the screen and put the next one up. "This is the model 200 springs. As you can see, the failures here are really quite similar except, on average, the springs lasted a little longer. We still have 31 of the high temperature ones fail."

Dan interrupts, "And in the same way, just losing their 'uuummmpphh'." He emphasizes the umph as he looks at Carl.

I continue, "Eight of the standard temperature ones, and only three of the low temperature springs." I pause after explaining.

Vic speaks up again. "What about the model 300 springs?"

I put the transparency up that shows it. "There were only 10 high temp failures, no standard temperature failures, and four low temperature failures."

Vic looks at Carl. "You look like you are deep in thought. What is your idea?"

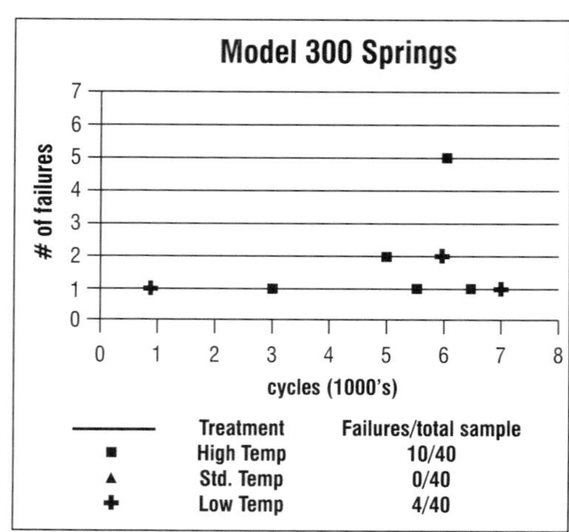

Carl is still looking at the data. He doesn't say anything for a few seconds then finally speaks. "Well, I think annealing is taking place with the smaller springs. And after talking to the engineers at HSM, the Springalloy vendor, I think I know why. The rods are pre-heated by passing through an oven that is attached to the front of the spring machines. Isn't that right Jess?" Jess nods and I realize he hasn't said anything yet. "The amount of pre-heating is determined by how fast the wire passes through the oven, AND" he emphasizes the "and", "the thickness of the material. In fact, when I explained our problem, he just about predicted what we see here. He was afraid that we were getting too close to the annealing temperature. That appears to be what is happening. What doesn't make sense is he said our failures would increase if our production temperature dropped too low, but it doesn't show that here."

Jess speaks up now. "Maybe it does. Dan, do you have those charts you and I did?"

Dan looks at Jess and nods. "Yeah, but they aren't fancy ones like Drew has, I don't know how to run the computer programs yet."

Amy encourages Dan. "Don't worry about that, if you want to learn to run them we will teach you. What charts are you talking about?"

Dan pulls some hand-drawn charts from a folder he has with him. "Last Wednesday when I was doing the tests, I found these papers that Jess had included in the test batches. Drew had told me how he did his first charts, so I thought I would try to do some too. These are simple ones, not fancy. All they show is how many springs broke during the wrap cycle while Jess was making these springs at the low temperature setting. While trying to make the spring it would often just snap the wire. Isn't that right Jess?"

"Yep." Jess agrees. "Lot of scrap. Just too brittle to wrap I guess. See, 43 of them snapped while making the 100 model springs and, as you can see, 29 while making the 200 model springs."

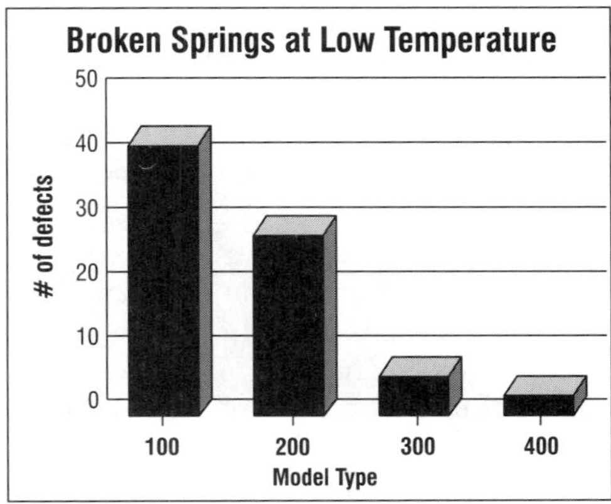

Everybody is leaning over the table to see the charts. It is a simple but neatly drawn Pareto chart showing failures during production. Carl and Vic both give out low whistles. "I KNEW it!" Carl exclaims. "If the material can take the wrap it has the strength to last forever."

Vic now speaks up. "But look at those numbers. To get 40 good samples Jess had to run over 80 pieces. I don't know about you but I don't want to propose a solution that reduces outside failures by 12 percent but doubles material scrap!"

Carl straightens up. "We don't have to. The HSM engineer said the idea is to use enough temperature to allow the wire to be bent but not so much that it anneals the material. It looks like we are in good shape with the 300 and 400 springs, we just need to find the right temperature for the 100 and 200 springs."

Amy then turns to Carl. "How do we go about finding the right temperature for the 100 and 200 models?"

"That's the great part," Carl replies, "WE don't have to. HSM will. This is their material. It's a special alloy they put together and they will determine the optimum processing temperature."

Jess looks suspicious. "Well, the temperature they already gave us doesn't work. What makes you think another will?"

Carl is excited about this now. "Listen, it just so happened I asked the HSM guy where they came up with the temperature that we use now. He said it was from the wire size we sent them. They produced wire that size and then determined the right temperature."

"OK then, why don't we have the right temperature for the smaller sizes?" Vic looks almost as confused as I feel.

Carl smiles. "Well, we only sent them the wire size for the model 300 springs. They didn't ask about any others even though they produce four different sizes. We both blew it."

Dan looks a little mad. "Who's fault is that? Those dummies should have realized that if we are getting four different wire sizes from them, that we are using all four sizes."

Amy tries to calm him. "Settle down. We should have sent all four specs. I suspect that it was just a lack of communication there, just like what happens here."

Carl explains, "That's exactly what it is. The production department is the one who got the order for the sizes and it was R&D that gave us the temperature specs. They didn't think to check with each other."

Amy adds, "Besides, at this point it doesn't matter whose fault it is, and it sounds like no one is really at fault. Let's just fix the problem and get on with our work." She sure is down to earth and job focused.

Amy then looks at Dan and Jess and adds, "And very good work you two. These charts were a missing link. If we had not had this data we could have made some bad mistakes. Thanks."

I give a thumbs up to Dan and he smiles. Then in his straightforward, almost demanding way he says, "Yeah, well if you want any more you better teach me to run the computer programs. I can't spend all my evenings writing up charts for you guys."

Amy almost laughs. "OK, we'll teach you. Drew, would you mind helping these guys? If you can't answer all their questions, then see me or Vic."

"Be glad to. I hope they can put up with a college boy." Dan, Jess and I all laugh and the others look at us funny. Carl mutters that it must be an inside joke.

For the rest of the meeting we determine what needs to be done. Carl says he would like to set up another experiment that would be a little more sophisticated. He wants to use a method called designed experiments. He says he will get another engineer who has a statistics minor to help him set it up and analyze it. He says the purpose of the experiment would be to identify the optimum temperature to run the model 100 and 200 springs.

Amy wants to know why that would be important if HSM determines the temperatures for us. Carl explains that even though they would determine the temperature for us, it would be good for us to better understand the material and the process, so we could more readily solve problems or make improvements in the future. He tells her he will prepare a brief justification for the experiment and that he is sure it wouldn't require any more resources than what was used on this project.

Amy agrees to his idea. Then Carl asks Jess to see if running the model 100 and 200 springs through the oven 20 percent faster will work out for the next two weeks until they get new specs for those models. Jess says he thinks it will be fine, but he will try it first thing in the morning and let Carl or Vic know if it won't work. If not, they can work from there. Jess suggests that he take 40 more springs at that faster rate and let Dan test them to see how they turn out. Everybody agrees that could be helpful. Then Dan asks what "college boy" is going to do. I guess he is worried about me.

Amy says that I am to write a report on the results of the tests and document the results. After that, I will start helping on a new housing design for the S-lines. Then she adds that I would be helping him and Jess learn how to run the computer so they could keep track of their own data and experiments.

Vic will be responsible for documenting the new process when it is determined, and making sure it replaces current production documentation. He will also continue to work with me and review my report before it is circulated and submitted to document control.

By the time the meeting is over, my mind is spinning. After we get back to our area, I ask Vic when all the stuff that was decided has to be done. "I had no idea so much testing and documenting was necessary."

Vic smiles. "It does seem pretty overwhelming doesn't it? Once you are sure that you have found the solution to a problem, then it is important to verify that solution and, even more important, that you make sure all the changes that are made are systemic. That means that the system is improved because of it and you don't just fix a symptom. Also, if you want the improvement to last, you must also remember to make it part of the system. That way if people change places, which they invariably will, then the improvements are still in place. We can also track the change and the reason for it, if necessary, in the future."

He invites me to sit down. "For example, your report. If somebody two years from now is working on a problem or improvement for the S-line solenoid springs, they will be able to go to document control and have them search for any work on that part. The computer database will lead them to your report, tests, and data. They can benefit from what you have already done and maybe it will help them get their work done faster or better. By the way, that work journal you are keeping . . . don't ever throw it away. It will have notes in it that you will be able to refer to when someone asks you about it, or if you have a task that is similar. Does that make sense?"

"Yeah, it does I think. But one thing. You said that you do all this when you are sure you have found the solution to the problem. How are we sure? It seems to me that we have hit a few points when we thought we had a solution, only to have something else come up to change our direction or add new information. I know I don't understand a lot of what happened today, but I'm just curious."

"Good question, Drew. I guess you could say that we aren't completely sure yet. That's why we are doing more tests. However, there are a few things that indicate we are on the right track. Let me ask a few questions. What did the results of Dan's test tell us about the failures of the high temperature model 100 and 200 springs?"

I think for a minute about what happened in the meeting. "Well, I'm not sure it told me very much, but that is when Carl brought up the correlation between the type of failure and the temperature."

"That's right. What he meant by that is that the material had obviously gotten too hot and lost its strength due to annealing. Right?"

I shrug. "I guess. I mean, I'm still not completely sure what annealing is, but I gather it was something to do with the temperature we were talking about the other day. Right?"

He nods. "Basically that's right. There are some other things that happen to the material but the strength characteristics we are after have been lost. Now the important point is that the data and the theory agreed on that point. The theory says that if we heat it too high and cool it slowly, we lose the strength that we want. As we saw from the data and the type of failure, that is what happened. Do you follow me?"

I nod. "I think I do, but it's obvious I had better pay attention when I take materials classes."

Vic smiles. "Not just your materials classes. Better pay attention in all of them. Including golf." He continues, "OK, do you remember when Carl got rather confused as to why the low temperature parts looked to be performing so well but something didn't seem to fit?"

I nod again.

"This was because something was missing for him. You may remember that Carl, and myself by the way, just thought, hey let's use the low temperature and go with it. But what didn't fit, fortunately for us, was Carl's understanding that a high temperature was needed for some reason. In other words, we still had this data point hanging out there that needed to be resolved. I think Jess and I were ready to ignore it. That is a strong urge, by the way, but you shouldn't do it. Carl wouldn't ignore it. His concern came from two sources—his understanding of materials, and his conversations with the HSM engineer, who had also indicated that high temperature was important. Do you remember?"

I nod again that I do.

"Well, when Dan and Jess showed the scrap results from the runs to produce the springs, it all fit. I haven't seen them, but I'll bet a month's wages that every one of those scrap pieces are fractured pieces that broke during wrapping. None will be just bent, they will all be snapped off because they were too brittle. That is why Amy called it the missing link. This Springalloy material needs to have enough temperature to make it producible but not so much that it anneals, or loses the long term strength we need in our springs."

He finishes talking and I feel drained. It makes sense to these guys. "So when the data and the theory match up, then you feel you got the answer?"

He goes to the white board he has in his office and writes:

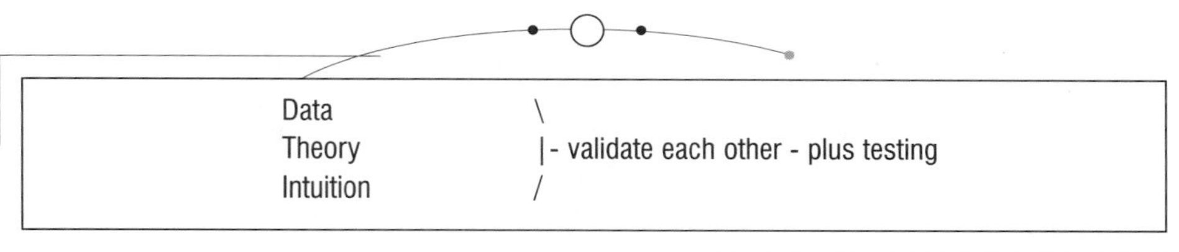

As he points to what he wrote on the board he explains. "When those all validate each other, AND further confirmation tests designed to test your conclusion agree, then you most likely have your answer and have reached the root cause of your problem."

I sit there overwhelmed. "Wow, I'm not sure I can do this. How do you learn all this?"

Vic just laughs. "Oh, you can do it. That is obvious to all of us. We have commented to each other on how good an engineer you will be in any type of engineering or engineering technology program. We each have our biases based on what we do and where we studied. But that choice is up to you."

I look at him. "What about you? What do you think I ought to study?"

He has a bit of a twinkle in his eye. "I came from an engineering technology program. They have a good one at State U. I think you ought to go that way." He smiles and adds, "Drew, you would do well in any of them. You need to determine where your greatest interests and talents are and go that way. If you have enjoyed your work here the last couple of months, then you would enjoy the field."

He goes back to his desk and sits down. "Talk to your dad about it over the break. Parents usually have a pretty good sense about this kind of stuff." He leans over and opens a file drawer. He shuffles through it for a while then takes out a couple of papers. As he hands them to me he says, "Here, these might be helpful in formatting and writing your report. Let me know when you are done, and I will review it for you. It's not a big hurry, but the sooner you get to it the better off you will be. Less chance of forgetting important information."

I had almost forgotten about the report. I thank him and go back to my desk. I start to look over the papers he gave me, but I have a hard time concentrating. I keep thinking about how this problem finally came together. Frankly, I think a lot about what Vic said. Everybody thinks I would make a good engineer! I don't know if all that is true, but it makes me feel good.

Vic peeks around the corner and to say goodbye. Says he's off to see his daughter swim in a swim meet tonight. I look at the clock and realize that it is already a quarter to five. I do have one thing on my mind I have to check out. On my way out I stop in the test shop, but I don't see what I am looking for there. I go over to the spring shop. It's pretty quiet but still open. When I walk over toward the spring machines that Jess runs, a guy running some other kind of machine asks me what I need. I tell him I'm looking for some stuff that Jess ran for me last week. He nods his head, points toward the spring machines and goes back to his work.

After looking around for awhile I finally see a box under his desk labeled, "Drew's springs - defects." It is the same kind of box in which the others were left. I rummage through it and find what I am looking for. I chuckle to myself. Vic was right. Almost all of them are completely broken. I do find one that is bent. I wonder for a minute if I should take it and show Vic. Maybe I could collect on that month's salary he was willing to wage. I decide against it though. What I have learned from him and the others is worth a lot more than a month's wage.

Chapter 17

I can't believe the Thanksgiving break is just next week. It has been about two weeks since we made the decisions on the S-line springs. The follow-up tests that Dan, Jess and I did to find a production temperature proved to be successful. The failures dropped as did the breakage during wrapping on the machine. In fact, when HSM finally got the specification temperature to us for making the springs for the 100 and 200 models, it turned out to be only a few degrees higher than what we were using.

Carl did his test also and came within just a few degrees of the temperature that HSM gave us. He was really pleased about that. He explained the designed experiment method he used. It was very interesting and looked pretty effective. It makes me anxious to get to the higher level classes so that I can learn a better experimental method. This whole experience has made me want to take more classes. I guess that's part of the neat thing about getting some experience. I can see how the classes will help me.

The report was harder to get done than I thought it would be. Between trying to get it written and staying in touch with Jess and Dan on the follow-up tests, I had a hard time sitting down and finishing it. Of course, knowing I have a deadline of right after Thanksgiving for my report for Dr. Stromberg's class helped provide some motivation. The chance to use the experience from work for my class has been great. It's made me think more about what Dr. Stromberg has taught. I left it on Amy's desk yesterday. There was a note on my desk when I came in that she wanted to see me about it. I hope it was OK, but I'll soon find out.

School has gone pretty well. I hit another study slump last week. I don't know why. I know what I need to do to study better. I need to do what I'm supposed to when I am supposed to. I just get lazy or something. I told Vic about my difficulty in trying to stick with the good study habits, and he said, "The best way to begin a new habit is to start." I told him he was as profound as Socrates. He then asked if I knew the best way to end a bad habit.

I said "What, stop it?" He laughed and complemented me on how quickly I learned things.

He did have some serious advice that was pretty good. He said if I wanted to make something a real part of what I do so that it becomes second nature, I should work on discipline as much as a specific habit. At first I wasn't sure what he meant. But he explained that all good habits are formed, and for that matter, bad ones broken, by being responsible and having discipline.

For example, he said if I am trying to form a habit of turning my homework in on time, but if even after a few weeks of doing so, I still struggle, it may be because I haven't adapted it to how I naturally work. And that applies not just to homework, but also in other areas of my life. He said that even though my homework might get turned in on time, at least most of the time, the habit is likely from fear of the penalty, not from discipline. I realize he might be right. I'm not late to work because I don't want to make anyone disappointed in me. I turn most of my assignments in on time because I will be penalized if I don't. He said discipline and taking responsibility are required to make it part of my life. Instead of making homework the only issue, I need to remove lateness from my mode of operation. Vic says that when I make punctuality the way I am, my homework will be in on time. It made some good sense, and I have actually made progress. The toughest part is getting to bed on time and up on time. Dorm life is notorious for late hours.

"Drew. . . Drew? Can I talk to you?" I realize it's Amy.

"Oh, I'm sorry, I was thinking about some stuff." I am embarrassed that I was daydreaming.

"That's OK," she answers. "Did you get my note? I'd like to talk to you about the memo and report you left for me. Come into my office."

I follow her into her office. Before she sits down she pulls out one of the guest chairs over by her desk. I get the feeling I'm about to learn something.

Amy is pretty no-nonsense and true to form she gets right to the point. "The stereotype among many engineering students is that most, if not all, of their time will be spent calculating, designing, or testing. In reality, that is the smaller part of their functions. I would estimate that engineering type people spend 50 to 70 percent of their time in some type of communication. In fact, as one moves into management, the estimate of time spent communicating probably increases to as much as 90 or 95 percent. Now, I'm not the best at oral communication, but I have tried to improve and I have really worked on my written skills. I want to give you some advice about writing. "First," she asks, "do you know why you need to write a memo and a report?"

Actually I'm not sure so I respond, "Not really, but Vic said I should write a memo to send with the report so people would know what they are getting."

"That is an important part of the reason." She takes a piece of memo paper from her desk and writes on it as she explains. "You want to communicate as simply and clearly as possible what you are sending and why. When I say 'what', I don't mean any of the details of the actual report except perhaps a sentence or two of the final result. Also, you want a short description of why you are sending it. Who it is going to obviously will appear in the header of the memo." She has written simply "what and why — clear and simple" on the memo sheet.

She continues. "In more formal situations, outside the company for example, you would use a cover letter, but the purpose is the same. Now let's use your memo to explain how to accomplish this purpose."

She stops for a moment to pull the report and memo I wrote from a file folder. When she opens it, I see my memo with a big X through the first paragraph.
"Your report is pretty good. The reason for the report is pretty well stated and the body and charts are good. The conclusion lacks clarity, but you can easily clean

that up. We need to help you get some economic analysis in the report so we can indicate the savings for the new process. That is critical information. Overall, the report is a little wordy, but I can show you how to improve that by using the memo as an example."

She puts the report aside, and I can see there are some comments on it, but I better concentrate on the memo.

"Let's look at the first paragraph. In fact, here. You read it."

I feel self-conscious reading it for her, but I go ahead.

> *"Advanced Component Technologies Corporation has been producing solenoids for their S-line series of solenoids for a number of years. These solenoids are used in a number of applications ranging from aircraft subassemblies to automotive applications. The solenoids for the S-line series are produced in four different sizes. Up until about two years ago the springs for these solenoids were made of high tensile steel and most of the springs were produced in ACTC's own spring shop. About two years ago, the material was changed to begin using a new alloy called Springalloy which is made by High Strengths Metals (HSM) who is the only vendor for this material."*

I stop reading and look up at Amy. She asks, "What do you know so far about the work you did on S-line springs and the results you got?"

I look back at the memo and pause for a moment. "Well, nothing yet. But I do talk about that down here." I point to the third paragraph.

"Drew, basically all written communications, whether they are memos, letters, or papers, have three parts. First is the introduction. The thesis, or the purpose of the memo or paper is stated in the introduction. Being at the beginning, it is the most reviewed part of the paper. One of the most common problems with the introduction is the lack of a clearly stated thesis. Can you see the purpose of the memo in the first paragraph?"

I see her point. "No I guess not, but I thought I would give some background first."

She nods as if to agree with me. "Background information is helpful and some of it is important. It would be better, however, to have it later in the body. In this case, it could be put in the paper, which you have done, and left out of the cover memo or letter."

I nod. I think I see her point. "So I should probably leave it out of this memo and just delete the first paragraph. Right?"

She smiles and simply says, "Read the second paragraph."

I'm thinking to myself, "Oh no, is it bad too?" but obey and begin reading.

> *"For the last two years, there have been a variety of defects occurring in the S-line solenoids, but the most serious ones have been spring failures. The greatest number of spring failures have occurred in the smaller models, that is the model 100's and 200's. Other errors have included cracked housings, burnt contacts and so forth. Because the spring failures in the small models seemed to be higher than they ought to be, it was felt that an analysis of the total number of failures and the percentage of each would justify further study. This study was conducted over the last 2 months at ACTC."*

I'm afraid to look at Amy because I think I know what she is thinking. So I say it for her. "I haven't said anything yet about the work we did on the S-line series, have I?"

She smiles and is kind to me. "Well, not really. But you did start to talk about the failures and failure types. In fact, your first two sentences of the third paragraph are great. Why don't you read them."

> *"Analysis of the failure types occurring in the S-line series shows spring failures to be very high in the Model 100 and 200 solenoids. A team of ACTC engineers and operators has determined the reason for the failures and found a solution which will increase thoughput and save money with no new equipment needed."*

"Now that's a beautiful introduction. As a manager, that's what I want to see. The important information is stated simply and clearly. Now it just needs to be right up front. Remember, the people who read this want to know what it is about, why it is important, what they will get out of reading it, and they want to know it quickly. In this memo you have great information but it is in the wrong place.

I think I see where she is coming from but I want to make sure. "Are you saying that the third paragraph should be first and the others after that?"

"No. I'm saying the last paragraph should be the only thing in your memo, except perhaps to add a sentence about the savings to be gained and that the attached report has details."

I slide the memo over in front of me and add one sentence. "Then would this be better?"

She reads it and says, "That is just exactly what I would like to see. Do you think you need anymore?"

I take the memo, tell her I'll be right back and make the changes on my computer. I return within five minutes, put the new memo on her desk and ask, "Is this what you would like to see?"

MEMO

TO: Amy

FROM: Drew, Dan and Jess

DATE: November 16th

RE: **Solution to S-line Spring Failures**

Analysis of the failure types occurring in the S-line model 100 and 200 solenoids shows spring failures to be very high. A team of ACTC engineers and operators has determined the reason for the failures to be process related. The attached report details the investigation into the failures and outlines a solution which will increase throughput and save money with no new equipment needed.

She reads it, then looks up and smiles. "Now compare that to your first one. Which one gives you the most information quickly? Also, which one makes you want to read the report?"

"I see your point." I tell her. "I think I know how the report can be improved as well. I'll make some corrections and get it back to you."

She hands me the report, then adds, "Drew, the ironic thing in all of this is that the better one becomes at communicating, the less time it will take to do it well. At least it will seem to take less time because the better you get, the more you will enjoy it. Being willing to develop and use good communication skills will open up many chances for individual growth and promotion. I have worked for ACTC for a number of years. I remember many times volunteering for numerous oral presentations that were required by our group, usually because other engineers did not want to do them. These presentations gave me exposure to different areas and people. Not only did I learn a tremendous amount about a variety of functions and problems, but the exposure to the other people and areas brought many opportunities that otherwise would not have come along. Some engineers complain that they should be judged on the quality of their work, not on their oral or written skills. The plain fact is oral and written skills are a part of their work, and if those skills are lacking, their evaluations will suffer."

She is good at what she does, that's for sure. "I hope this was helpful," she says.

"This was great. I know writing is important and you gave me as many good idea as I have gotten all semester. Thanks for taking time. I'll get busy on the report."

It took longer than I thought it would to finish the report. In fact, it was two days later before I gave it back to Amy. When I apologized for how long I took she didn't look at all surprised and just said, "You'll get better. Thanks for the good work. I'm looking forward to reading it."

As I walk out I hear her call after me, "Oh Drew, have a good Thanksgiving."

I have been so busy finishing up homework, tests and the report for the S-line work that I had almost forgotten that the day after tomorrow I get to go home for Thanksgiving break. Dad said if I couldn't find a ride home then he would come and get me, but I've got a ride home with Shaun Stevens and his dad. I came up with them at the beginning of the semester. I wonder what kind of puzzle Mr. Stevens will have for me this time. I'm looking forward to the break. I know I'll want to get back to school, but I'm looking forward to seeing how Dad is.

Vic has me working on some of the details of the new housing changes for the larger S-line models. I don't have as much time to do what he wanted because I am finishing up the spring reports. I really want to get that stuff done.

It's also been about a week since I've been down to the shop to see Dan and Jess. It has been a lot of fun getting to know them better. It didn't take long to teach them all I know about the word processing and charting programs I use on the computer. They caught on very quickly, but they both have used computers before, so it wasn't a huge step. I need to get down to see them before I leave for Thanksgiving.

The report is done, and it's in Amy's hands. I decide to leave because I need to get some more study in for two tests tomorrow — physics, which is going better since my movie mistake of 3 or 4 weeks ago, and art history. I like parts of that class, but some parts just don't keep my interest. But that's OK. If I liked every part of every class I take, people would think I'm not a true freshman.

Chapter 18

"Well son, how has school been?" I've been home for a few hours now, and it's good to sit down and relax. When I first got home, Mom and Dad wanted to show me the new houses that are being built where I used to play pick-up soccer. Some rich guy, Dad says. But we're home now and just sitting around. The best news was the report about Dad's cancer. Things look great with no signs of any more cancer. He says he feels good and work is going well.

"It's been good, Dad. I've enjoyed it and I think I'll do pretty well this semester."

"All A's?" He looks hopeful.

"I don't think all A's. Art history class is going well but I think I'll have a B in that one. The others I think I may have all A's in or at least close."

He nods approvingly. "Good." He kinda mutters under his breath. "How about that job you have? Has that worked out OK?"

"Oh, that's been great! Let me tell you about a project I got to work on there."

I spend the next twenty minutes telling him about the S-line stuff. Of course I have to include the time I spent in Stan Hall's office talking to him. I even tell him about the headlight story with Stu. He seems pretty impressed.

"So, is my son going into engineering?" Without letting me say anything he yells, "Hey honey, we have an engineering genius in the house!"

"I have enjoyed the introduction class, and the things we have been doing at work are great." I was going to say something about making good money but I've started to see that's not all there is to it.

Mom walks to the door, and just as she does, my little brother Joey runs into the room. He is very excited. "Drew, are you going to learn to drive trains? Will you take me for a ride? Please! Please! Mom, tell him he *has* to take me for a ride!"

While Dad and I laugh, Mom puts her arm around him and says, "I think Drew may be the kind of engineer that makes things, not drives trains, honey."

He looks disappointed. "I want Drew to drive trains, not make things." He sulks off back to his toys.

Mom sits down too, "Engineering huh? What kind?"

"Well, I'm not exactly sure yet."

Before I can say anything more, Dad re-enters the conversation. "Well, what kind of engineering have you been thinking of?"

"Well, there are about six different fields of engineering and engineering technology at State." I'm enjoying my time to explain what I know. "The ones they have at State are manufacturing, mechanical, civil, electrical, and chemical. I know a little about them because we learned about them in the Intro to Engineering and Engineering Technology class I am taking."

Dad leans back like he's ready to listen for awhile. "What are the differences anyway?" he asks. "I've heard about all of them, but don't know too much about them."

I tell him what I know and then he asks about the differences between engineering and engineering technology. I use the information I remember from a lecture early in the semester by Dr. Stromberg to explain it to him. He seems very interested.

"Any of the people at your work from the engineering technology programs?" He asks.

"Yes. In fact, my boss at work graduated from the engineering technology program. The vice-president of quality at ACTC was also an engineering technology major. He is a pretty good manager, I hear. I haven't talked to him much though."

Mom breaks in. "You guys about ready to eat?"

That really gets my attention. "You bet, real food. I can't wait."

As we start dinner, Mom asks, "How did you get along with your roommate, I think you said his name was Stu? Does that stand for Stewart?"

"No, Stu comes from his last name. His real name is James, but his dad has the same first name, so Stu goes by Stu to tell them apart." "Stu and I have gotten along well. He's a pretty outgoing guy, but really nice. We have even been able to help each other with homework sometimes."

"That's great." Mom seems happy about that. "What other people have you met?" She asks with the "any girls in your life?" look but I tease her by ignoring it.

"Well yes, there's Troy. We are in the same physics class, and he's been a great guy to be around. Then there's Russell who is also a good friend."

Sis interrupts, "Mom wants to know if you have a girlfriend, you dummy."

Dad laughs and I smile. "Nope. No girlfriends. . . but I have been out a few times, Mom. Lest you worry that I'm not meeting any women, Troy and I met two girls at a goal setting class at the beginning of the semester, Christy and Kellie. I have been out with Christy a few times and Troy and I have double dated with them also. They are neat girls. Christy is pretty and fun. She just loves everything about life and has a neat handle on important things. In fact, she's interested in the same programs I am. She would do great."

"That's wonderful, Drew. Have you been out with anyone else?"

"Oh, two or three others once or twice but not many. I also just have been pretty busy between work and school."

Sis speaks up again. "During a phone call a few weeks ago you said your roommate had a sister that you liked too."

I give her the "be quiet, you pest" look but she ignores it. Mom asks about her.

"Stu's younger sister is a senior in high school this year. She won't be at SU until next year. I met her when Stu's family brought him to school. I went with him to their house for dinner. They only live about a half-hour from campus. Katie, that's Stu's sister, is a little quiet but just a neat person. She is easy to talk to and really doesn't overreact to things. I think she's pretty level-headed. I hope to get to know her better next year. She plans on studying engineering also."

We chat some more about girls, school, and what's happened at home. After dinner, Mom has to take Sis to a student council meeting of some sort, and Joey heads back to his room to play. That leaves Dad and me alone.

Dad is quiet for a minute. "So, have you learned anything, Drew? At school I mean."

I think for a minute. "Yeah, Dad. I have already learned a lot." I know Dad well enough to know he is talking about more than just the stuff from classes.

"I've learned that discipline and responsibility are important characteristics." I think I see a small smile on his lips. I go ahead. "I've learned that all the classes I take can help me in whatever I choose to study and can help me in life. Of course I need to take them seriously." I think the smile is getting broader. "And I have learned that it's not as easy as I thought it would be to break some of the bad habits

I got into and to make good ones. But," I add quickly, "I am doing it. Sometimes it's been a little tough. But," I repeat, "I am doing it."

Dad is smiling now. He couldn't hold it in any longer. "You got a couple more growth rings, huh?"

I look at him funny. "Growth rings??"

His smile gets bigger. "You know a tree is growing because it gets growth rings. Each ring stands for more experience, more knowledge, more strength. And sometimes the rings show the tree has had some tough times. But those times often mean a tougher tree and stronger wood."

He reaches into his pocket, pulls out his wallet and takes a small card about the size of a business card from it. "Here Drew, take a look at this. This is my personal creed for learning and growing throughout my life. Some might call it a plan for life-long learning. I call it maintaining personal development."

Maintaining Personal Development

— Live a life of personal integrity.
(Honor promises and expectations)
— Have a rich private life.
(Reading, music, hobbies)
— Spend time with nature and its wonders.
— Continue to learn and grow.
— Give service in your work and community.
— Strengthen your family and keep close contact.
— Practice continuous improvement.

I read them one by one. They're pretty impressive. I have always admired Dad for his positive outlook on life and the way he always seems to be improving. I can see how these things have been incorporated into his life. After reading them I start to hand it back but he stops me.

"You keep it Drew. I have another copy. Maybe it will be helpful to you."

Just then we hear that Mom is home again. She has Aunt Mary with her. She wants to see me too.

As I walk out of the room with Dad I say. "Yeah Dad, I guess you're right. It hasn't even been one semester, but I guess I have another growth ring or two."

Part Two

Drew's Notes

Section 1 — History of Engineering and Technology

Drew Barnes — Reflections on History of Engineering and Technology

We discussed the fact that our time and technology are the best the world has ever seen with respect to the products, services, and conveniences available. Because of this, there can be a tendency to view past civilizations as less intelligent than ours; however, it is not true. Deep-thinking, innovative, and creative people have lived in every time. We are fortunate to be living in the greatest technological age ever, with more opportunity for education and learning than ever before, thanks to developments and discoveries over many centuries. There is much to learn from those who preceded us, including a deep appreciation of the advances they made. Even in technology, where we are obviously more advanced, we can marvel at their insight.

Regardless of the time in history, the ingenuity of mankind is nothing short of amazing. The needs of society, such as food, shelter, protection, or water, have been tightly linked to the engineering and technological developments of the time. In fact, in many cases the era is defined by the dominant technical discovery of the time, such as the stone age, the bronze age, the plastic age, or the information age, or by the great engineering and technology feats of the time, such as the pyramids of Egypt and the road system of Rome. Throughout history, the activities now commonly described as engineering and technology have had as much or more impact on civilizations, their survival, and standard of living than any other single aspect of civilization.

The great thinker and scientist, Sir Issac Newton, said "If I have been able to see farther it is because I have stood on the shoulders of giants." As modern engineers and engineering technologists, we also stand on the shoulders of giants, and our opportunities are greater because of others. Some of the great inventors and thinkers were of the enlightenment when modern science was developed, and their contributions are briefly mentioned.

Another point made in the class was about "triggers." Triggers are key inventions that cause a dramatic change in the development of society. According to James Burke, the first trigger for technological advancement in recorded history was the plow[1]. All innovations are, to some degree, a trigger because they promote change and advance knowledge. Some inventions are especially influential because of the number of people or the time span they affect. The printing press was one example.

Also a few examples of some key inventions gave us a chance to discuss in class the nature of ingenuity, as well as the effects on society of the different kinds of inventions.

The notes that follow are the handouts from the discussion on the history of engineering and technology. Even the short discussion we had gives me a better appreciation of the work that has gone on for centuries and that this profession has been, is now, and will always be a very valuable one to society.

[1]Burke, James, *Connections,* Little, Brown and Company, 1978

Assignments: Divide students into groups of 5 to 7 people. Use the quiz to illustrate certain points about engineering and technology. After each question, give each group one minute to answer. Award the top scoring group a prize.

The Quiz (answers are in the instructor's manual)

1. George Crum, a Native American, invented which of the following?
   A. The potato chip
   B. The graphite pencil
   C. Plastic
   D. Glass
   E. The staple

2. James Burke identified eight innovations which he described as being the most influential in history. Which of the following is one of the eight?
   A. Numerical controlled machine tools
   B. Injection molding of plastics
   C. The mass production line system
   D. Iron smelting
   E. Statistics

3. Who said that genius and invention are "99% perspiration and 1% inspiration"?
   A. Lincoln
   B. Morse
   C. Newton
   D. Calvin and Hobbes
   E. Edison

4. According to James Burke, the first "trigger" that caused a major increase in the standard of living and social development for people in the Nile river valley was:
   A. The dam
   B. Wheat
   C. Irrigation
   D. The plow
   E. Iron

5. Where was the first formal engineering curriculum in the U.S. established?
   A. West Point
   B. Deep Springs College in Dire, Nevada
   C. Brigham Young University
   D. Harvard
   E. State University of New York

6. How much manufacturing involves the production of electronic components or devices?
   A. 15%
   B. 28%

   C. 35%
   D. 50%
   E. 75%

7. Who said, "We need as many engineers as possible. As there is a lack of them, invite to this study persons of about 18 years who have already studied the necessary sciences. Relieve the parents of taxes and grant [them] sufficient means"?
   A. Emperor Constantine
   B. Ben Franklin
   C. George Washington
   D. Herbert Hoover
   E. Winston Churchill

8. The inventor of the mass production (then called Division of Labor) method was:
   A. Leonardo da Vinci
   B. Adam Smith
   C. Eli Whitney
   D. Samuel Colt
   E. Thomas Edison

9. During the 1960's, more money was spent on what consumer product than on the entire Apollo space program?
   A. Hair grease for men
   B. Mascara for women
   C. Mouse traps
   D. Car wax
   E. Roller skates

10. If the automobile industry had made the same progress as the electronics industry, a Mercedes Benz would cost _____ and get _____ miles per gallon.
   A. Under $100 & 10,000
   B. Under $50 & 50,000
   C. Under $20 & 100,000
   D. Under $10 & 500,000
   E. Under $2 & 1,000,000

11. The term "engineer" was first used in:
   A. 200 A.D.
   B. 1000 A.D.
   C. 1700 A.D.
   D. 1850 A.D.
   E. In 1923

Assignment: (an example is given in the instructors' manual) Research a device or invention from before the year 1800. Submit a neat sketch of the item and write one half to one page describing the purpose of the item, the principle(s) on which it works, and why it was selected.

History of Engineering and Technology Overview

Developments Throughout History
- Ancient times
- Middle Ages
- Renaissance
- Industrial Revolution
- Modern age

The Trigger Effect
- Facilitator of advancement
- Major triggers throughout history

Selected Inventions
- Printing press
- Mass production system
- Plastics
- Telephone

History of Engineering and Technology Developments Throughout History

Ancient Times (4000 BC–400 AD)

Mesopotamians

- copper then bronze then iron tools
- terraced temples
- ships using oars and sails, trade and colonization
- schematic symbols and writing, primitive math, 60-based numbers
- clay used for writing medium
- sophisticated wheel and axle systems
- surveying and measuring tools
- irrigation systems, roadways and structures
- astronomy
- use of the number "0"

Egyptians (3000–600 BC)

- enormous pyramids
- chief engineers were royalty figures
- extensive roads, canals, flood control systems
- mechanical devices such as levers, ramps, etc.
- schematic symbols and writing, primitive math
- papyrus used for writing medium

History of Engineering and Technology Developments Throughout History

Greeks and Romans (600 BC–400 AD)

Greeks known for **intellectual advancements**
— modern alphabet
— study of nature and classification
— logic systems
— astronomy and celestial models
— use of parchment

Romans known for building **practical devices**
— arches and domes
— sophisticated system of roads
— concrete
— aqueducts and irrigation
— improved ships
— sewer, plumbing, and municipal water systems
— heated bath systems

History of Engineering and Technology
Developments Throughout History

Middle Ages (400 AD–1400 AD)

Period of slowed growth but still some bright spots
Term "engineer" used commonly
- — an "ingenitor" designed and built war machines
- — inventors were referred to as "ingenitors"

European
- — windmills and waterwheels
- — bladed plow
- — stirrup

Chinese
- — developed gunpowder
- — animal collar (harness animal energy)
- — paper

Arabs
- — advancements in chemistry
- — a systematic classification for substances
- — optics and lenses
- — discovered cinnabar (a pigment)
- — produced alum (a fixer and hardener)

History of Engineering and Technology Developments Throughout History

Renaissance (1400–1700 AD)

- Period of enlightenment
  - —printing press
  - —scientific method
- Some great thinkers came from this time
  - — Leonardo da Vinci
    - gifted thinker and great artist
    - detailed steam engine, helicopter, submarine, camera
  - — Francis Bacon
    - described the process known as the scientific method
  - — Galileo
    - gifted artist and writer
    - corrected many ancient beliefs
    - telescope
  - — Isaac Newton
    - laws of gravitation and developed modern calculus
    - worked with light and colors
  - — Robert Hooke
    - experimented with the elastic properties of materials
    - determined stress is proportional to strain
  - — Robert Boyle
    - defined the expansion and compression of gasses
    - determined gas pressure, volume, and temperature
      - directly related
  - — Anton Lavoisier
    - modern chemical method
    - quantitative science

Page 5 of 8

History of Engineering and Technology
Developments Throughout History

Industrial Revolution (1700–1900 AD)

- Textile machines
- Harnessing of electrical power
- Basis for communications technology
- Mass production method
- Educational revolution with increased number of people involved in science and engineering

Thomas Newcomen

— invention and application of steam engines

— provided method to pump water from mines in England

Modern Age (1900–present)

- Half life of technical knowledge and skills at $1\frac{1}{2}$ to 4 years
- Computer technology
- Jet aircraft
- Satellite technology (Sputnik, etc.)
- Space flight
- Materials advancements
- Medical advancements
- Atomic power

History of Engineering and Technology
The Trigger Effect

Triggers

Accelerators of advancement

Stimulates a major paradigm shift caused by key inventions and advancements

Spawns other inventions and advancements

May not be the best known of inventions

Every innovation is a trigger of change to some degree

Some major historical trigger inventions are

—plow

—printing press

—steam engine

—electricity

—telephone

—computer

—plastics

—nuclear power

—Internet

Page 7 of 8

History of Engineering and Technology
Selected Inventions

Trigger Type Innovations of Various Ages

Printing press

- — was to its time what the computer was to the I900's
- — provided dramatic increase in dissemination of information
- — basis for increased education and communication for discovery and inquiry

Mass production system

- — first described by Adam Smith in 1776 in "The Wealth of Nations"
- — significantly improved uniformity of product
- — provided huge increase in production volume
- — concerns voiced by Smith included lack of mental involvement by workers

Plastics

- — lowers price of products
- — very versatile in their applications
- — can replace many other materials
- — easy to manufacture and shape

Telephone

- — provided a means to communicate immediately anwhere
- — combining of many inventions and technologies

Secton 2— Study and Advisement

Drew Barnes — Reflections on Study Skills and Succeeding at College

The class on study skills and being successful at college identified a number of things I can do to increase my success here. Some things I was aware of, but some things I didn't know before and will make my study more effective as I use them.

The first part of the class was on time management—that is, making and keeping a schedule. I have done better here at school, at least so far, but that might be because it is still new to me. The real test of these things is maintaining them, not starting them.

There were a few points suggested for using time well that I hadn't thought of before. For example, I had not really considered the odd times, or short time breaks, and how they could be used more effectively. Time standing in line, or even longer times like between classes, are opportunities to get some quick tasks done that I haven't utilized well before.

Then the four time robbers that were talked about are all problems for me to one degree or another. As much as anything, the point that caught my attention was the idea the teacher brought out about continuing to make progress. I don't remember exactly how it was said, but the idea was that the difference between success and failure is making one more attempt. The successful person keeps trying until they are successful. The person who fails is the one who gives up. Troy, who was in the class with me, showed me a proverb he had taped in the front of his planner. It said, "That which we persist in doing becomes easier; not because the nature of the task has changed, but that our ability to do it has increased." I am coming to understand that persistence, hard work, and dedication are more important than intelligence.

Other points included advice on using a textbook effectively. There were a lot of things suggested here that I had not considered. We were advised to keep as many textbooks as we could. The teacher said many more students regret selling back their books than ever regret keeping them. There were also some great suggestions on how to underline and mark a textbook. For example, too much marking can be as bad, or worse in some cases, than no marking at all. The practice of marking only important topics is the best approach. Using consistent marking methods is important, too. I'm going to practice these techniques in all my classes I think, but especially in physics. I want to learn to highlight only the most important and fundamental things.

The information on taking notes was a good review of many things I learned in high school but have not been using well. The one thing that did stick out is the point that one hour of study right after class is better than two hours later. I could save time by doing that, and with just a few minor adjustments in my schedule, I think I could get in at least one hour of study right after all but one of my classes.

The last points in the class were about using resources like the professor, office hours, advisors, and study groups. Some of the advantages mentioned for visiting regularly with the professor or advisor are points I had not considered. It is pretty clear though that keeping regular contact will be a real advantage. We each should be responsible for ourselves and not expect others to do things for us, but sometimes some good advice goes a long way. I think I'm like a lot of students, and I feel a little intimidated by professors. But they are real people, and they want to help. The teacher admitted we may run across some who are not as open to visits as others, but not to lose hope and realize that they are busy too and have a lot of things to do. All in all, the things I learned will help out a lot. I guess I had better set some goals now to use what I learned.

The notes, handouts, and assignments from the class follow.

Assignment: Study and Time Management Complete a weekly schedule similar to the example shown. Challenge each student to follow it to the letter for one week, then write a 1–2 page paper describing the experience. The paper should describe successes, things learned, how the schedule could be modified to be more helpful, and challenges that were faced during the week.

Assignment: Advising Each student is to work out a specific graduation plan, then have an advisor sign it and submit it as the assignment. Students who have not yet declared a major should choose one they may be interested in and do the same thing.

DAILY SCHEDULE - Fall Semester					
Time/day	**Monday**	**Tuesday**	**Wednesday**	**Thursday**	**Friday**
7:00					
8:00					
9:00					
10:00					
11:00					
12:00					
1:00					
2:00					
3:00					
4:00					
5:00					
6:00					
7:00					
8:00					
9:00					
10:00					
11:00					

Study Skills and Succeeding at College
Overview

Factors That Will Contribute to Success at College

Be organized
- — planning through using a schedule
- — hints for using time well

Use textbooks and lectures effectively
- — reading texts
- — marking texts
- — taking notes

Take advantage of resources available
- — working with advisors and teachers
- — group study sessions

Page 1 of 6

Study and Advisement
Factors for Success—Organization

Use a Weekly Schedule

Schedule fixed commitments (classes, work, appointments, etc.)
Make a to-do list and prioritize (check rank of priorities every day)
Do high priorities first
Schedule flexible commitments such as studying, eating, sleeping, etc.
Leave some time open for emergencies
FOLLOW the schedule

Other Hints for Using Time Well

Utilize odd times that are easily wasted, ie: the hour between classes
Use short breaks (5–10 min) to avoid burn out
Carry something to work on while eating alone, in lines, etc.
Set realistic priorities
Beware of the 4 time robbers

—laziness
—sidetracks
—procrastination
—daydreaming

Study and Advisement
Factors for Success—Using a Textbook Effectively

Reading a Textbook

Preview the chapter (get an idea of what's going on)
Ask a question you think will be answered in the first section
Read (try to answer the question)
Recite from memory what was learned
Repeat process for all sections
Review points from each section until you have learned the chapter

Marking a Textbook

Write useful notes in the margins (helps you remember and understand)
— if you don't want to mark up the book, use small slips of paper (like Post-it® notes) and leave them in the pages of the book
Underline only important parts of sentences, not whole sentences
Underline after reading
Use ═══ for main points and ___ for supporting points
Don't underline too much—just very important material
Use other symbols (*or | lines) in margins, but sparingly

Study and Advisement
Factors for Success—Using Lectures Effectively

Taking Notes

Before the lecture
— read or skim assigned reading material before class
— identify unfamiliar terms/concepts (listen for them in class)
— review previous lecture notes

During the lecture
— label notes by date, subject, chapter, topic
— listen and concentrate; what's the topic and point?
— avoid distractions (where or by whom you sit, etc.)
— watch for visual cues of important points (words, posture, gestures)
— listen more than you write (write neatly)
— ask questions if you don't understand

After the lecture
— review and edit notes as soon as possible
— correlate with text, handouts, etc.
— fill in gaps, and ask questions of a classmate or the teacher
— review periodically

Hint: 1 hour of study right after class is worth more than 2 hours a few days later.

Page 4 of 6

Study and Advisement
Factors for Success—Taking Advantage of Resources

Suggestions For Working With Advisors

Visit them at least once a semester to:
— review courses and progress in program
— check course offerings and possible changes

Let them know of class conflicts and overlaps
— they might have hints to help your schedule and avoid other conflicts
— informing them will help other students

Advantages of regular visits
— they will know you in order to provide letters of recommendation
— they can be aware of situations and can help in case of hardship
— they may know experts who can help in
tutoring
jobs
co-op opportunities
graduate work

If you feel a part of the department and program and you know what is going on, you will do better and be happier. Your suggestions and ideas will also help the department improve.

Study and Advisement
Factors for Success—Taking Advantage of Resources

Other Suggestions For Working With Teachers

- Remember they are real people and they love what they do
- Go see them when you have concerns and questions; they want to help
- Know when office hours are; use them or seek a special appointment time
- Attend study/help sessions with the teacher whenever possible

Advantages of Study Groups

- Build confidence and help prepare for difficult tests and assignments
- Get perspectives from fellow students (helps learning and retention)
- Expand the material the teacher has given

Your Part to a Successful Study Group

- Be on time, review material before coming, keep commitments
- Don't expect study groups to make up for your lack of attention in class

Note: Study groups are hard work but most are very successful.

Page 6 of 6

Secton 3 — Goal Setting

Drew Barnes — Reflections on Goal Setting and Personal Development

The few pages that follow are some of the notes and outlines from the class on goal setting and personal development. I learned that there are a number of factors that will contribute to my success at college, and that I have control over most of them. I can even improve them with a little knowledge, practice, and discipline (this means hard work). The teacher pointed out that contrary to some popular attitudes, other people, other things, and even our past do not dictate our ability to be successful and happy. Development of the habits and characteristics in this class will not only help me be a better student, but will be beneficial in all areas of my life.

Setting and achieving goals was the first topic. Few things have quite the same impact on accomplishing objectives as setting and working toward challenging goals. It is possible to live without goals, but life will mean more and challenges will be easier to deal with if we have positive, challenging, and worthwhile goals. So my goal is to set and achieve some worthwhile goals.

Then the point was made that I am in college to succeed, not to fail. Therefore, I need to be sure to set myself up to be successful. Some of the things I can do to ensure my success and increase my satisfaction with my college experience were mentioned, but the focus was on methods and characteristics of personal development. The teacher reminded us that this whole thing, both goal setting and personal development, is a process that takes time. In fact, we work on it our whole lives. This quote was given: "Self-examination is an important part of goal setting and commitment, and might be one of the most difficult steps. In the process of self-examination, however, one should not become obsessed with solving all the problems at one time."

Then we learned how important a good attitude is. There is a fairly popular adage that sums up the importance of attitude. It goes, "It doesn't matter what happens to you in this life, it is how you deal with it." Learning to deal with whatever happens to come along is a key attribute to success in any endeavor—school, work, or personal life. Those who tackle problems, and work through them with an attitude of learning, hope, and persistence will experience less stress and greater satisfaction than those who just lament the challenge and simply try to ignore it or hope it will go away by itself. There are ways to improving your attitude and ability to tackle problems with a positive attitude.

The last thought was the importance of always developing myself and my personal characteristics. I learned the Ben Franklin method. Franklin worked on a characteristic each week that he wanted to improve. After 13 weeks, one week for each characteristic, he would start over again, therefore always improving.

As part of this point on personal development, fundamental characteristics of discipline, responsibility, and integrity were talked about. Developing these three, the teacher said, will prove very valuable no matter what we do or where we work. These are characteristics of life, not just of a field of study.

Though it was kind of a pep talk, the class reminded me of some pretty important things. The other thing that impressed me is that skills and characteristics that can help me be a good engineer are important in all areas of my life. I had not thought much about the transportability of these skills.

Assignment: Goal Setting Students were asked to determine and set a major goal with measurable sub-goals that will be associated with their learning this semester, write it on a sheet of paper, and submit it to the teacher. This is similar to what is described in the class on goal setting. The goals are kept until the end of the semester, when a follow-up activity will ask for a report on the progress of the goals and what was learned.

Goal Setting and Personal Development
Guidelines for Goal Setting
Overview

Goals Should Be Your Own

- More likely to accomplish your own than someone else's goals
- Some goals should be directed toward personal development

Goals Should Be Clear, Specific, and Concrete

- Easier to commit to clearly stated goals
- Achieving specific goals helps develop discipline
- Easier to know when you have accomplished them
- Sub-goals, tasks, and rewards must support the main goal

Have a Specific Timetable for Accomplishment

- Assign target dates for completion of goals
- Sense of success and accomplishment

Goal Setting and Personal Development
Guidelines for Goal Setting

Goals Should Be Written ("A goal not written is only a wish")

Clarifies your thinking and increases commitment

Reminds you of exactly what it is you are going to accomplish

Share Goals With Others Who Can Encourage You

Increases commitment and provides support to the goal

Share with someone who cares, has insight, or experience

Goals Should Be Realistic but Still High

No stretching means no growth

Real value of a goal is what happens in the process as well as the result

Have to change habits, thoughts, and activities, but worth it in long run

Set Times Regularly to Review and Revise Goals

Review frequency depends on the time frame of the goal

Weekly goals reviewed daily, monthly reviewed weekly, etc.

Goal Setting and Personal Development
Personal Development

Learning About and Improving Yourself

Talents and strengths

— require work to keep and develop
— inspire
— breed confidence
— bring satisfaction and happiness
— uplift others

Faults and weaknesses

— discourage and depress
— slow progress and personal growth
— bring frustration

The idea is to increase talents and eliminate faults. Like a garden, you nurture the good so the bad has no room to grow.

Goal Setting and Personal Development
Personal Development

Fostering a Positive Attitude

"The longer I live, the more I realize the impact of attitude on life. Attitude, to me, is more important than facts. It is more important than the past, than education, than money, than failures, than successes. It is more important than appearance, giftedness, or skill. It will make or break a company . . . a church . . . a home. The remarkable thing is we have a choice every day regarding the attitude we will embrace for that day. We cannot change our past . . . We cannot change the fact that people will act in a certain way. We cannot change the inevitable. The only thing we can do is play the one string we have, and that is attitude . . . I am convinced that life is 10% what happens to me and 90% how I react to it."

Charles Swindoll

Developing a Positive Attitude

Look at challenges with a longer term perspective
— ask "What can I learn from this?"

Don't be a slave to a poor disposition or attitude
— you do have a choice in the matter

Consider how lucky you are
— happy people are grateful for what they already have

Identify and emulate positive characteristics in other people

Page 4 of 5

Goal Setting and Personal Development
Personal Development

Developing and Following a Personal Development Improvement Plan

Ben Franklin method
- — identified 13 personal characteristics he wanted to improve
- — concentrated on 1 characteristic each week
- — repeat after the 13th week
- — continue indefinitely

Set and accomplish goals
- — follow goal setting process
- — make goals with respect to personal attributes
- — accomplishing goals itself improves attitude and discipline

Build Fundamental Personal Characteristics

Discipline
- — critical to accomplishing anything of significance
- — persistence and hard work are critical to success

Responsibility
- — fosters self-confidence
- — necessary for leadership

Integrity
- — foundation to sound self-esteem
- — ensures a good reputation

Secton 4 — Problem Solving and Design

Drew Barnes — Reflections on Problem Solving and the Design Process

This is great stuff. Getting into some of the things that I will be using throughout my career was exciting. Besides, problem solving is just plain fun. Like a detective solving a mystery, it's a chance to discover. That's a big part of why I want to be in engineering and technology.

We learned that there are at least two ways to gain understanding in terms of problem solving—cognition and reasoning. Both are used, but one may be more prevalent than another, depending on the problem. Cognition is a process of awareness or a gut feeling. Part of this is intuitive, but it is enhanced by practice and experience. This sixth sense, as some call it, is very valuable in solving problems.

Reasoning is the ability to think and is improved through education and experience. It involves the understanding and interpretation of facts.

We also learned a process for solving problems. I have already used it a few times. The teacher told us that the process might vary slightly from problem to problem, but the general process works well. We were told more than once during this class to be sure to seek the root cause of a problem rather than the symptoms. Symptoms will go away if we get at the root of the problem, but the reverse is not true. We were also told to be sure to not act just to get something done. Act when necessary, but don't forget to think carefully and clearly.

We learned a few hints to improve problem solving. There are things that can be done at each step in the process. They will help ensure that root causes are discovered and that the solutions we come up with are good ones.

After problem solving, we learned about the design process. In some ways, the design process is like the problem solving process. Many of the steps are similar, and the things that make the design process valid are some of the same things that ensure good problem solving. A critical part of the design process is making sure that the needs of the customer are met, and it is the best measure of how well we have done our jobs.

Another thing that I really didn't understand until it was explained is the value of constraints in design. I guess I always felt that constraints would limit my freedom and make the design process (or even the problem solving process) much more difficult. It is true that it makes it harder, but the very nature of constraints is why we have opportunity for design. It was pointed out that constraints bring validity to our solutions and design, and our solutions and designs are better with constraints. In other words, if there were no constraints, solutions would not be as good.

In addition if there were no problems, then a large part of the reason for our work would disappear.

The next few pages are notes and outlines from the instruction on problem solving and the design process.

Assignment: Design Problem #1

Part 1 - Week 1

Each person should design a paper airplane for a flying competition. The object is to make an airplane that travels exactly 32 feet. The airplane can be launched by hand or you can design a special launching mechanism. In the competition each person will get three flight attempts. If the airplane travels further or shorter than 32 feet, points will be lost. Your final score will be based on the average of the three attempts, so your airplane should be as consistent as possible.

The rules are as follows:

- The sheet of paper your airplane is made from can be no larger than 11″ × 17″.
- Tape, glue, weights, etc. are all permissible.
- The airplane must fly through the air, not slide across the ground. However, if it slides on the ground after landing, this will be included in the final distance.
- Engines, motors, or radio controlled devices are not permitted.
- The aircraft must resemble a airplane and act like a airplane. No "paper wads" etc. will be permitted.

Remember to keep your design simple. Keep good documentation of your design and how it was produced.

Part 2

Get together with an assigned group of 3 to 4 people and decide, as a group, on a final design that your group will use for both the airplane and the means of launching it. It may be one of the individual designs or it could be a combination of the best ideas in the group. The group will only have one airplane, however. As in Step 1, you should use good documentation on how to make the airplane and the means of launching it. Again, you will be evaluated based on flight consistency and closeness to 32 feet.

Part 3

Provide a set of instructions that another group must follow to the letter as to how to construct your group's airplane and launching mechanism. No communication with the other group other than the written material should be allowed. You will be evaluated on how close the results of their flights come to your own group's results.

Problem Solving and Design Overview

Process for Solving Problems
- Cognition
- Reasoning
- Procedure

Aids to Good Problem Solving
- Key to better success at each step

The Design Process
- Goals
- Requirements
- Methodology

Aids to Good Design
- Constraints
- Creativity
- Customer

Problem Solving and Design
Problem Solving

Process for Solving Problems

Cognition

— coming to know, including through awareness and judgement

Reasoning

— the ability to comprehend and infer through orderly, rational thinking

Procedure for Problem Solving

— define the problem clearly

— obtain information

— analyze the information

— determine a solution

— implement and test the solution

Remember to think carefully. Don't act for the sake of acting

Problem Solving and Design
Problem Solving

Aids to Good Problem Solving

In problem definition

- — learn to observe; problems send out signals; experience and observation will help you be sensitive to these signals
- — watch for piles of paper, product, waste, etc; these are signs of problems in action

In obtaining information

- — use data gathering methods such as Statistical Process Control
- — communicate with other people/departments; the perspectives of others are very valuable

In analyzing information

- — use basic charting techniques such as histograms, flowcharts, scatter plots, time series charts, etc.
- — look for patterns, trends, links, etc.

In determining a solution

- — recognize and practice order; problems occur as things fall out of proper order
- — find root causes; symptoms come from causes; only resolving causes will solve problems

In implementation and testing the solution

- — involve all relevant people/organizations
- — carefully plan implementation including how results will be evaluated
- — document procedures, tests, and results carefully

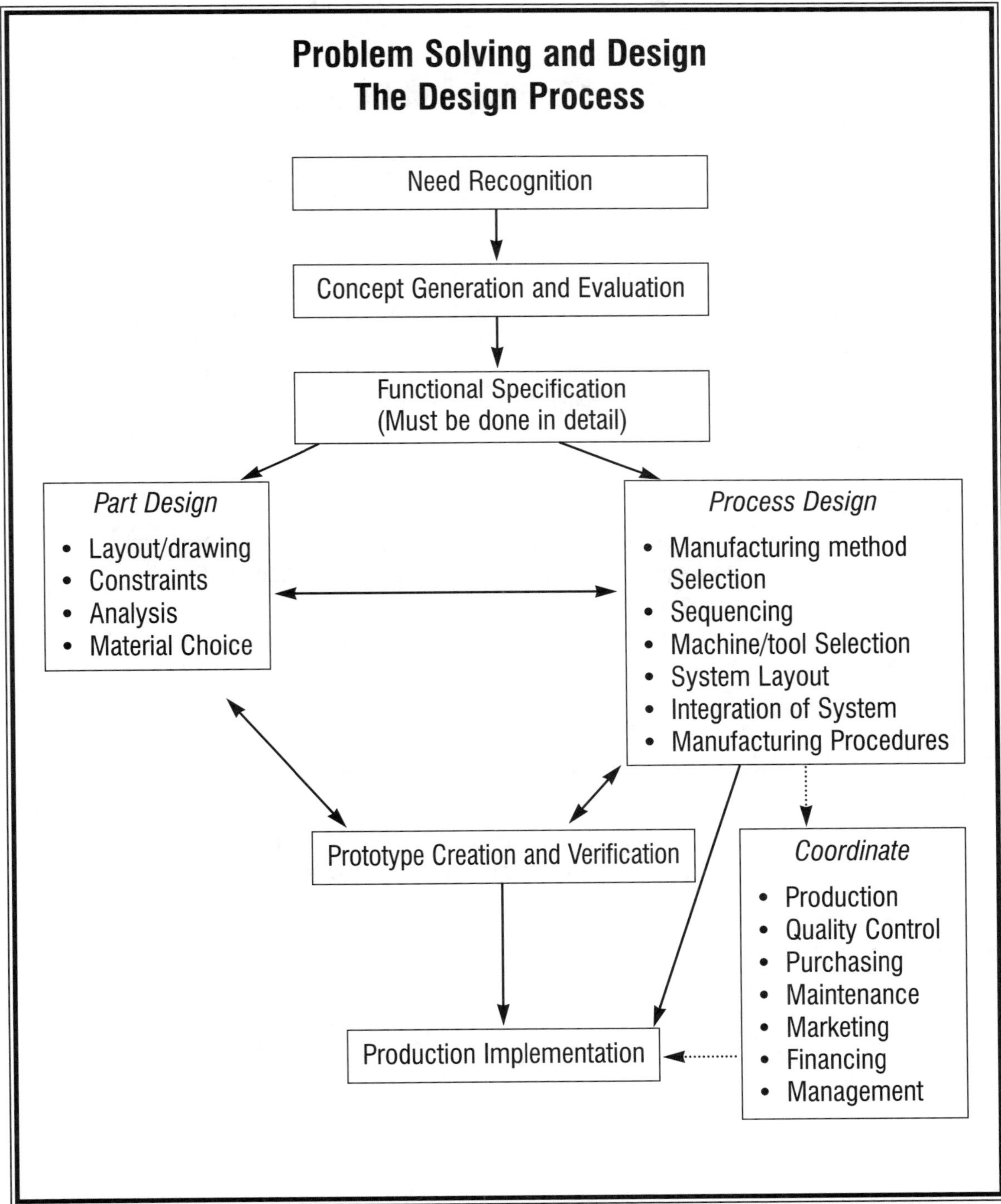
Problem Solving and Design
The Design Process
Need Recognition
Concept Generation and Evaluation
Functional Specification
(Must be done in detail)
Part Design
• Layout/drawing
• Constraints
• Analysis
• Material Choice
Process Design
• Manufacturing method Selection
• Sequencing
• Machine/tool Selection
• System Layout
• Integration of System
• Manufacturing Procedures
Prototype Creation and Verification
Coordinate
• Production
• Quality Control
• Purchasing
• Maintenance
• Marketing
• Financing
• Management
Production Implementation

Problem Solving and Design

Aids to Good Design

Constraints

— describe accurately and clearly
— review in the course of the process to ensure all are defined
— make sure both internal and external constraints are defined
— constraints are critical to good design and problem solving; should be viewed as a benefit, not an imposition

Creativity

— encourage creative brainstorming early in the concept stage
— encourage innovative thinking of current methods in the design stage
— encourage inventive thinking in the implementation process

Customer

— every product, process, and service has a customer
— identify and exceed customer expectations
— don't expect the customer to know all of his or her needs; visit the customer and use your knowledge and insight to understand their needs

Page 5 of 5

Secton 5 — Practice of Engineering Technology

Drew Barnes — Reflections on the Practice of Engineering and Technology

This was another useful and interesting lecture because it described, in a general way, what I would be doing with my career and the benefits of this field of study. I learned that engineers and engineering technologists are creative problem solvers who work with analytical and visual tools and methods to accomplish their tasks. They are integrators; that is, they work with a variety of people in numerous functions to design and produce better products and to serve society. They are hardworking and enjoy being able to contribute to others and their work. This description fits what I would like to do.

We learned that the study of engineering is closely related to science and is based on scientific principles. Science strives to explain natural phenomena through observation, experimentation and hypothesis. Any engineering task or achievement is based on proven natural or physical laws that have been described through science, whether they be by laws of motion, thermodynamics, quantum physics, etc. These fundamental principles define the way the natural and physical world works. John Prados, in speaking of the link between science and engineering, said, "The goal of scientific activity is knowledge, an understanding of the physical universe in which we live. The goal of engineering is to create a device, system, or process that will satisfy a human need."[1]

We also talked about the unique aspects of engineering and engineering technology. The engineer and the engineering technologist are quite similar but have different inclinations and expertise to apply to problems and ideas. The background of the engineer includes more math and physics (thus the inclination and ability to develop sets of rules by analytical modeling using science and math), whereas the engineering technologist has more experience in the actual application of the process, operations, and methods. The engineering technologist also has an inclination and ability to use existing rule sets and apply them in innovative ways to solve problems and improve products and processes through experience and experiment. The differences are emphasized more in academic circles, whereas most industrial functions and organizations call them engineers and treat them both as technical professionals.

We studied the attribute of being good engineers and engineering technologists as well as the benefits of a career in engineering and technology. Some of the attributes of being a good engineer or technologist were really interesting. For example, the importance of keeping good notes and records is something I hadn't thought of before.

In terms of the benefits of studying engineering and technology, the teacher explained that it provides an excellent way for you to learn to think well, develop

[1]Prados, John W., *Journal of Engineering Education,* January 1997, pl.

intellectually, and contribute to society. These same skills can be of benefit to society. Even though the technical skills we obtain might not be directly applicable to the social or environmental ills of the day, our ability to analyze problems, generate ideas, and implement and evaluate solutions is greatly needed and wanted. I had not considered the benefits outside myself and my company before. Some great thoughts that reinforce my feelings that this is a good place for me.

The teacher also talked about some of the tools we would be using and introduced us to spreadsheets. Spreadsheets can be used to do some simple calculations and display output for some math problems. Examples were demonstrated and assignments given.

Assignment: See the assignments and activities in the problem solving and design section. The set of problems completed in that section apply to this section also.

Assignment: Using a spreadsheet package, solve the following problems.

Spreadsheet Problem #1 (Same Problem as Statistics Problem #8)

A turning process produces 9 shafts with the following diameters:

2.3 in., 2.27 in., 2.25 in., 2.31 in., 2.32 in., 2.30 in., 2.28 in., 2.27 in., 2.27 in.

Using a spreadsheet program such as Microsoft Excel, calculate the (a) average diameter, (b) median diameter, and (c) standard deviation.

Spreadsheet Problem #2

A test engineer has found several problems that are occurring in an injection molded housing for an instrument panel. The following problems occurred during the last year:

Problem	# of Occurrences
Too much flash around the edges of holes	437
Warpage of the part	125
Burnt parts	52
Surface imperfections	22
Cracks where ejector pins hit the mold	11
Internal Voids	7

Using a spreadsheet such as Microsoft Excel, (a) make a histogram of the occurrences, and (b) make a pie chart showing the percentages of each occurrence.

Practice of Engineering and Technology Overview

Definition of Engineering and Technology
- General description of tasks
- Founding Principles

Differentiation of Science, Engineering, and Technology
- Nature of the work of science, engineering, and technology
- Value of all areas

Attributes of Successful Engineers and Technologists
- Record keeping
- Continual learning
- Proficient in problem solving and design

Benefits of Engineering and Technology Careers
- Personal
- Career

Practice of Engineering and Technology
Definition of Engineering and Technology

General Description of Tasks

Creative problem solving
Use analytical and visual tools and methods to accomplish tasks
Integrators working with a variety of people and functions
Design and produce goods and services to serve society and people

Defining Principles

Engineering is the application of science and mathematics by which the properties of matter and the sources of energy in nature are made useful to people in structures, machines, products, systems, and processes. (*Webster's New Collegiate Dictionary*)

Key Aspects

Application of science and mathematics
Properties of *matter* and *energy*
Made *useful* to people

Page 2 of 5

Practice of Engineering and Technology
Differentiation of Science, Engineering, and Technology

Science
- Describes natural and physical laws
- Recognition and formulation of problems
- Collection of data through observation and experimentation
- Formulation and testing of hypothesis

Engineering
- Develops set of rules to apply science, mathematics and the properties of matter and energy
- Design structures, machines, products and processes

Engineering Technology
- Innovative application of engineering rules to provide products and services for human use
- Implements design rules and principles to ensure quality product

Technician
- Uses existing tools and techniques to accomplish task
- Expert in specific machine or sets of machines

Value of All Areas
- Require different talents and education, but all are important to society and organizations

Practice of Engineering and Technology
Attributes of Successful Engineers and Technologists

Keep a Lab/Workbook—Example: Thomas Edison

- Documents discoveries for patenting or credit purposes
- Record flashes or insight or inspiration
- Provide legal protection in case of litigation
- Reliable way to recall problems, progress, solutions and issues for further work
- Writing clarifies thinking

Continually Learning

- Seek out and learn from general education subjects (history, music, literature, religion, etc.)
- Promotes value to company and society
- Desire to learn new things is a critical characteristic
- Use graduate education, professional societies, company education classes and programs

Proficient in Problem Solving and Design

- Iterative process promotes learning and improvement
- Requires expertise in analysis, prediction and formulation
- Appreciate practical side of trial and error work of engineering
- Engineering is dynamic, as is problem solving and design

Practice of Engineering and Technology
Benefits of Engineering and Technology Careers

Personal, Career, and Society

Job satisfaction
Employment security
Challenging work
Constantly improving products and processes
Uses new technology
Understand how things work
Learn to think well
Provides skills to contribute to other areas
Understand cause and effect relationships
Ability to analyze problems
Provides skills to avoid ill-founded ideas
Improves standard of living for society
Help improve the standard of thought in society (see below)

"Science and technology have immeasurably enriched our material lives. If we are to realize the immense potential of a society living in harmony with the systems and artifacts it has created, we must learn . . . to use science and technology to enrich our intellectual lives."

James Burke

Secton 6— Teams and Teamwork

Drew Barnes — Reflections on Teams and Teamwork

We learned in the class that there is a 100 percent chance that we will be working in teams when we are in the industry. Of course, the fact that we will be working for a company, a team itself, I suppose, makes this topic very important.

First we talked about some of the different kinds of teams, and the way teams are used. These all made sense, but the point that I had not really thought of before was the idea that teams could actually promote continuity in the work of an organization when used effectively. Since people leave organizations for various reasons, and with the constant development and technological change that occurs, teams can help provide an organization with continuity. I know it is something that is mentioned in sports teams. For example, we hear that SU's basketball team has six seniors returning, three of them starters. This, they say, will help the team maintain some coherence from last year to this year. This also applies to teams in other organizations. People know how others on the team work and think. This allows easier communication and better decisions. However, too much continuity can also cause lack of innovation. The ideas and perspective of new team members will add a fresh view of the world.

We also talked about the different kinds of teams. One of the types that was described was the informal team. These are the people who get together without being organized by a formal authority. The teacher told of a group of people who were graduates of the same university (a small alumni group) in the first company he worked for that used to get together each month to have lunch and talk about their work. They began to share information with each other that helped them all. This informal team became a very important part of the information gathering group for these people.

Some fundamental aspects of the team process were mentioned as well. Things such as the need for a clear objective, that trust is critical, and that developing individual skills helps the team. We also talked about what a team needs to successfully operate and how to appropriately use teams. Teams can't, and shouldn't, do everything.

We then learned about the four stages of team development. The most impressive part of this discussion to me was the idea that all groups and teams go through some kind of conflict or disagreement stage. What was interesting is that this stage is critical to reaching the highest level of team performance. The different views and perspectives that come out in this stage are necessary to provide the basis for the highest, or performing, stage of team development. The very differences that people have conflict over are the differences that people begin to appreciate at the performing stage of development. The point was made that the same happens in all groups, even families and marriages. The point to remember is that conflicts can be resolved.

The last thing we did was review some different roles in teams and groups. Anyone can play all of the roles in a group, they just do it different ways depending on their personality and talents. We also learned about negative roles. These are people who play roles that stop progress. We were cautioned however, to not judge someone too quickly. For example, someone who is a good reality tester, or devil's advocate, might be helping the group make a good and valid decision, but might be seen to be blocking progress.

The pages that follow are the notes and outlines from the discussion on teams and teamwork.

Activity: A team activity is conducted in class. Instructor has materials and instructions.

Teams and Groups
Overview

Advantages and Purposes for Teams and Groups
- Information
- Decisions and ownership
- Continuity
- Communication

Types of Teams and How Teams Function
- Informal
- Ad hoc
- Cross-disciplinary
- Temporary
- Team processes
- Appropriate use of teams

Group/Team Development
- Stages of development
- Resolving conflict
- Achieving excellence

Group Roles and Individual Responsibilities
- Roles in groups/teams
- Expectations of group members
- Link to the group purpose

Teams and Groups
Advantages and Purposes for Teams and Groups

Information

More minds working on the problem
More information available
Better chance of getting all the facts
Leads to a better decision

Decisions and Ownership

More people in the process
More people committed to the decision
Better chance for support of solutions to the problem
Better chance of the implementation being successful

Continuity

More people involved in the problem
Greater awareness of the issues by broader group
Less need to "re-invent the wheel" in later efforts
Better chance of consistency in follow-up activities

Communication

Face to face contact can increase trust
Cross-disciplinary functions involved together improves understanding
Shared information enhances quality of the system
Interpersonal skills improved through joint work

Teams and Groups
Types of Teams and How Teams Function

Informal
- Not assigned by authority
- Groups who gather due to common interest or by happenstance
- Often very tightknit organization
- Sometimes leads to formal organization
- Can meet a social as well as functional need

Ad Hoc
- Team formed for a short-term purpose
- Formed from immediately available resources
- Usually have a specific problem to address

Cross-Disciplinary
- Representation from multiple organizations and functions
- Usually long term goals and purposes
- Typical of many industry-based teams
- Promotes system improvements

Team Processes
- Develop clear objectives and goals
- Basis of trust and openness
- Use appropriate leadership
- Practice good inter-group relations
- Encourage individual development

Appropriate Use of Teams
- Must be given appropriate authority
- Management must understand teams cannot fix management problems
- Make sure teams do not work against each other
- Ensure the problem is one a team should be formed for

Teams and Groups
Group/Team Development

Stages of Development

Forming
- — introduction
- — basic rules established
- — objectives agreed on

Storming
- — characterized by conflict and disagreement
- — caused by different opinions, personalities, and expectations
- — 50% of teams fall at this point
- — it is possible for any team to move past this

Norming
- — tolerance of others views, opinions
- — good working relationship
- — team is becoming productive
- — conflicts resolved or at least "agree to disagree"

Performing
- — appreciation of others views, opinions, talents
- — characterized by very high productivity
- — close knit relationship
- — high level of synergy
- — an environment where talent can flourish

Resolving conflict

Many methods and techniques can be used
Most important is for members to know resolution is possible
Good leader can be very helpful

Achieving excellence

High personal satisfaction can come from high performing teams
Excellent team skills breed success in other aspects of personal life

Teams and Groups
Group Roles and Individual Responsibilities

Positive Roles in Groups and Teams

Leader/coordinator
- — keeps order
- — provides direction

Reality tester
- — provides critical analysis of idea (devil's advocate)

Harmonizer
- — attempts to resolve disagreement

Summarizer
- — provides ability to pull together ideas/discussions

Researcher
- — provides data and information for analysis

Gatekeeper
- — attempts to keep communication open

Negative or Blocking Roles

Aggressor
- — stops progress through intimidation

Dominator
- — monopolizes time, resources

Comedian
- — impedes progress through comical and immature behavior

Page 5 of 5

Secton 7 — Mathematics and Statistics

Drew Barnes — Reflections on Mathematics and Statistics

This lecture on math as the language of engineering and technology really opened my eyes. Before I had just thought of math as something I had to use to figure out problems. But it is more basic than that. Math is like a language that we use to communicate information in science, engineering, and technology more effectively than any other language.

The elements of this language—mathematics, visual representations, and formulae—provide for communication across disciplines and specialities. Successful engineers and technologists will develop and maintain skills in all of these areas. Depending on the discipline you work in and the type of work you choose, different levels of expertise will be required and varying levels of use in these elements of engineering language will be necessary. But knowledge in each will be important, no matter which discipline you choose. As with any language, it is important to learn proper principles of organization and use. Communications in engineering and technology, due to the sometimes complex nature of the problem and solution, require logic, clarity, and proper procedure.

We learned how each of the major branches of mathematics provide different focus and strength in this whole thing. Algebra is an advancement of arithmetic that uses symbols and operators to represent unknown numbers and sets. These symbols and operators are used to do manipulations quickly and accurately. Just as English sentence structure has a form and order, such as noun, verb, and adjective, so does an algebraic sentence have a structure, namely variables (often symbolized by letters of the Greek or English alphabet, for example X), expressions, and relationships.

Geometry is another branch of mathematics that describes rules and laws associated with the relationships between various configurations of lines, angles, circles, polygons, and other shapes. Geometry proves these relationships in a very logical way. Therefore, it is also a method and system of thinking. Practice at this can help our ability to think and solve problems, which we need to become very good at anyway.

Trigonometry is based on the study of triangles. We talked abut how triangles appear in all of nature and the world. Not that things are necessarily made of a lot of triangles but that they describe a lot of relationships that are then easily measured or modeled in triangle form.

Calculus is the branch of mathematics that studies the rate of change of functions using what are called limits and integrals. Differential calculus deals with calculating maxima and minima of functions and instantaneous rate of changes of functions. Integral calculus calculates areas and volumes bounded by curves and surfaces, and can find the length of curves. Real movement and time are also often best calculated using calculus.

Statistics is simply the study of using a lot of numbers, called data, to find out if the numbers are important, and how important they are. The rules of statistics allow us to make generalizations based upon data from a few samples because we understand, or we assume, that the entire group follows the standard rules. (I was amazed at how many circumstances followed normal patterns.)

Though it sounds a little complex, it is obvious how these all are important in engineering and technology. I had better understand as much as possible. Being able to at least communicate well in math and statistics will be a great benefit to me in engineering and technology.

Assignment:

1. A man is standing on a street and looking northeast at a tree on a perpendicular street as in **Figure 1.1.** If the man could walk 40 ft. to the east, then turn north and walk 30 ft. to the north and reach the tree, then how far would he walk if he just went straight to the tree?

North

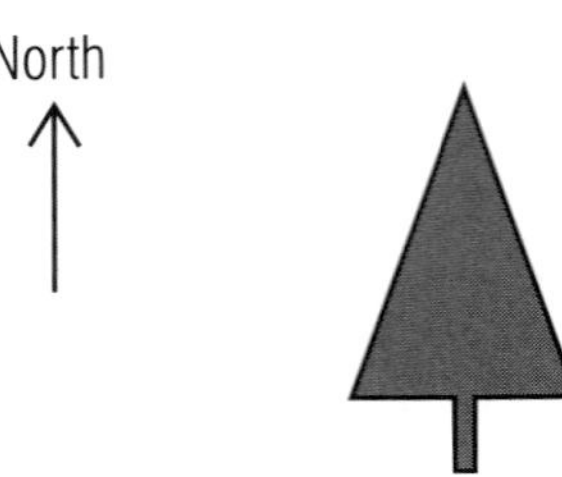

Figure 1.1

2. A robot is programmed to move from one corner of a room to another as shown in **Figure 2.1.** By facing the opposite corner it can move in a straight line to the other corner in a distance of 23.32 ft. at an angle of 59.03° from the wall to its right side. What are the dimensions of the room?

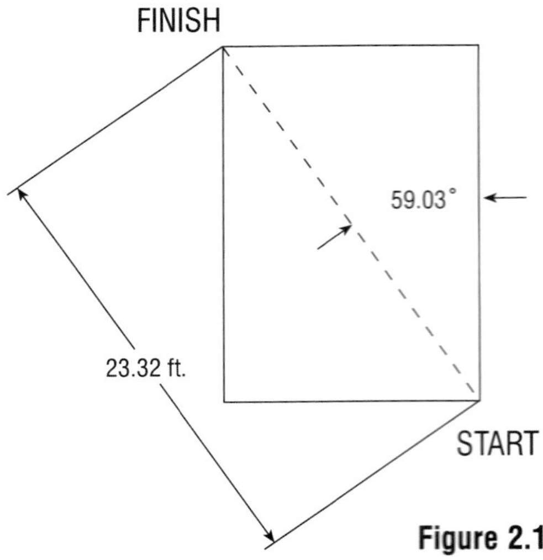

Figure 2.1

Use trigonometric functions to solve for a and b as follows:

$$\sin A = \frac{\text{opposite}}{\text{hypotenuse}} = \frac{a}{c}$$

$$a = c(\sin A) = 23.32(\sin 59.03) = 20\text{ft.}$$

$$\cos A = \frac{\text{adjacent}}{\text{hypotenuse}} = \frac{b}{c}$$

$$b = c(\cos A) = 23.32(\cos 59.03) = 12\text{ft.}$$

3. A toy manufactuer is designing a push toy to be used by small children as shown in **Figure 3.1.** The toy's handle will be held at the end, at a height approximately 2 ft. off the ground and an angle 25° from vertical. How long should the handle be?

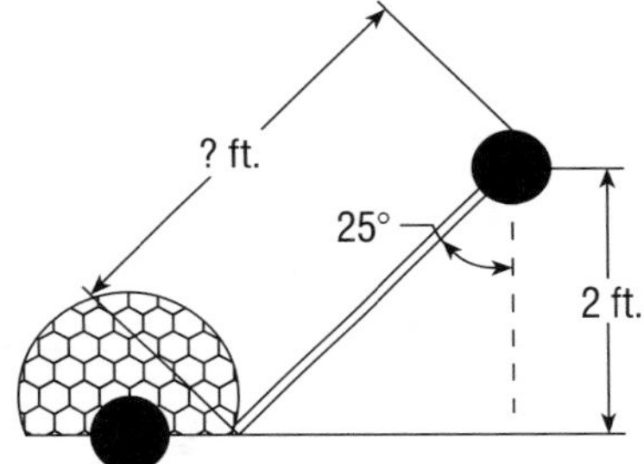

Figure 3.1

4. Silicon disks with a 4-inch diamter have been manufactured to make integrated circuits. What is the circumference and area of the disk?

5. Calculate the area of the I-beam's end view shown in **Figure 5.1.**

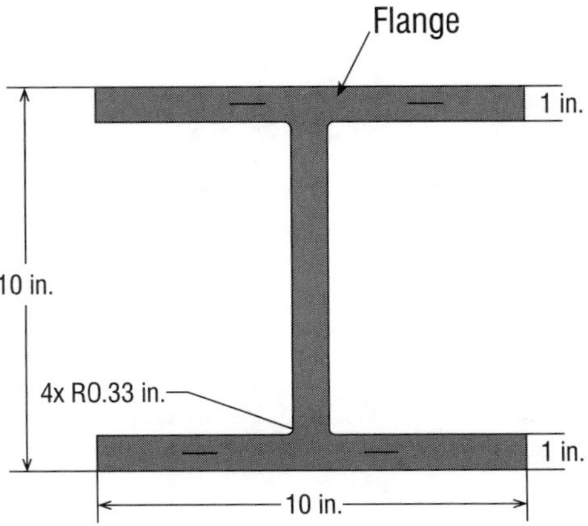

Figure 5.1

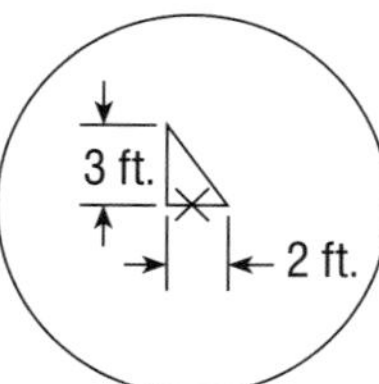

6. A 25 ft. tall, 15 ft. diameter water tank is shown in **Figure 6.1.** A solid triangular-shaped rod extends through the whole height of the tank. What is the volume of the tank? What is the weight of the water that would fill the tank? (Density of water is 0.0361 lb/in.^{3})

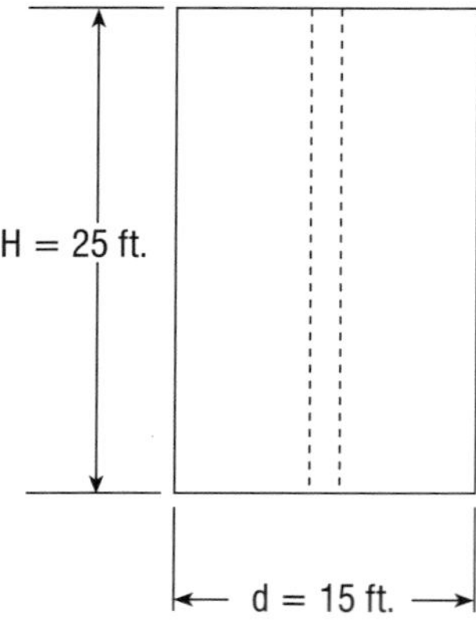

Figure 6.1

7. During a test of a new sports car, the following data were recorded. Graph the data using basic graphing principles.

Time (seconds)	*Speed (miles per hour)*
1	18
2	33
3	47
4	60
5	72
6	83
7	93
8	102
9	110
10	117

8. A turning process produces 9 shafts with the following diameters: 2.3 in., 2.27 in., 2.25 in., 2.31 in., 2.32 in., 2.30 in., 2.28 in., 2.27 in., 2.27 in.

   Calculate the (a) average diameter, (b) median diameter, and (c) standard deviation.

Mathematics and Statistics
Overview

Mathematics: The Language of Engineering and Technology

- Algebra
- Trigonometry
- Geometry
- Calculus

Statistics

- Data analysis
- Data production
- Statistical inference
- Value of statistics
- Numbers vs. data

Mathematics and Statistics
Mathematics

The Language of Engineering and Technology

- Mathematics allows for clear communication independent of any other language
- Describes analytical subjects more accurately than any other language
- The language of choice among the scientific and engineering communities
- Provides exact operational methods and symbols

Algebra

- Uses the power of the equal sign to establish relationships
- Provides the ability to solve for an unknown quantity
- Teaches a host of powerful manipulations
- Expresses equations graphically—for example:
  - — line
  - — curve
  - — parabola

- Represents math concepts in spatial terms
- Is fundamental to other mathematical branches
- The science of reduction and cancellation

Page 2 of 5

Mathematics and Statistics
Mathematics

Geometry

Earliest mathematical science
A system of logical thinking
Deductive logic—helpful skill for problem solving
Teaches relationships that exist in the world of geometrical figures
Has a rich and long history
Serves as the basis for other fields
Newton's laws are based on fundamentals of Euclidian geometry
Describes multi-dimensions

Trigonometry

Developed a system of relationships that have very practical use such as the following:
— surveying
— measuring height
— mapping

Based on the study of triangles
— triangles are a common natural occurrence

See much of trigonometry in life

— sines and cosines, in waves, electronics, atomic movement, etc.

Page 3 of 5

Mathematics and Statistics
Mathematics

Calculus

Is based on the concepts of

— limit - the reduction of a real or theoretical entity to the infinitesimally smallest quantity that still reflects the nature of the entity

— differentiation - the mathematical process used to find the limit

— integral - the summation of a real or theoretical entity within certain bounds, such that the error between the calculated and the actual summation is infinitesimally small

Calculus

Can represent space and time

Provides for study beyond finite or static to the continuous

Critical background for study of higher mathematics

Page 4 of 5

Mathematics and Statistics
Statistics

Statistics

Is the science of understanding information from numerical data using

- — analysis
- — graphs
- — charts
- — probability
- — inference

Practical Parts of Statistics Are

Data analysis

- — methods for organizing and describing data

Data production

- — methods for generating data in order to answer specific questions and ensure accuracy

Statistical inference

- — methods and rules for drawing conclusions about data including estimates on the validity of the data

Value of statistics

- — gaining understanding from a sample of a large population of data

Numbers vs. data

- — data are numbers with a purpose
- — larger amounts of the data provide the basis for statistical analysis

Statistics allow us to assess whether differences in data are meaningful.

Secton 8— Engineering Economics

Drew Barnes — Reflections on Engineering Economics

Economics was more interesting than I thought, and it seems to be important for us to know. We were taught that money and the use of it is one of the most common constraints we will find in our work in engineering and technology. Some knowledge of how the economic side works will help in understanding financial statements generated by a company, managing budgets for projects or helping groups communicate better with financial departments and people. It will also help us make better decisions regarding project or product payback, and prepare us for management. There is no way to become an expert in one session, but this gave us some basic knowledge.

In our work in engineering we will need to make decisions about equipment to improve a process; justify a new plant, line, or product; or even expend funds for training. We will need to determine the costs, when the funds will be spent, calculate what is called the rate of return, and find out how long it will take to make the investment payback. Most companies have a set amount that a project will need to return in order to be considered for approval.

There are a number of terms used in this area. Many of them make sense and are fairly clear. A lot of things will take more study. For now, I got the idea of some basic concepts, such as depreciation, benefit/cost ratio, capital, and return on investment.

The concept of time value of money was pretty interesting. The value of money is determined by when you receive it. For example, a dollar you receive today is worth more than one you receive tomorrow. If you received the dollar today and put it in the bank, you would have the dollar plus the day's interest. If you wait until tomorrow to get the dollar then you only have the dollar. Over a period of time, this can equal quite a sum.

Taking a simple example on the time values of money, we used the equation for future value, $Fn=P(1+i)^n$. We were to figure how much total money would be in the account if \$1000 were placed in a savings account for 3 years at 6% interest.

$$F_3=\$1000\ (1+.06)^3 = 1000\ (1.191) = \$1191$$

Therefore, you earn \$191 just by having your money in the account. This means that each dollar is worth about \$1.19 three years from now at the interest shown. In this example we calculated interest compounded annually. Today, interest is compounded more frequently such as quarterly, monthly, daily, or even continuously. There are more examples in my notes.

The point was made that in our work we will likely need to determine what is the value of a process improvement or a new product, machine, or something like that for our company. If the improvement cannot compete, for example, with the return to be made by simply having the company put the money in a savings account, then why spend the money on the improvement? This makes it important for me to understand the value of money, the value of a proposed improvement, how long it will take to gain a full return on an investment in a machine, or other capital investment, and what to expect as a value on the return.

Assignment: Determine the costs and benefits, in economic terms only, of purchasing a car for $8000 and using it to commute to school on a daily basis versus using the bus. The situation is that you live about 5 miles from campus and walk 4 blocks to the bus stop. Account for all costs in both situations and compare the cost. Calculate the total costs over four years of school for both modes of transportation. Then determine the future value of money saved on gasoline, insurance, maintenance, etc., if the bus method was used instead of the car. Assume that this money was invested at a given rate.

Assignment: Prepare and conduct a two week money log in which you log every expenditure made, and the purpose, time, and place of the expenditure. At the end of the time period, evaluate the expenditures and comment on the relative worth, both monetary and intrinsic, of the expenditures.

Engineering Economics
Overview

Terminology of Engineering Economics
- Return on investment
- Capital
- Depreciation
- Benefit/cost ratio
- Inflation and deflation

Concept of Time Value of Money
- Current value and future value
- Principal and interest
- Equation for calculating simple interest

Justification for New Equipment
- New product
- Regulatory/safety
- Increase production/improve throughput
- Savings on labor
- Improve quality

Examples
- Time value of money
- Future value
- Equipment justification
- Payback

Engineering Economics
Terminology of Engineering Economics

Return On Investment
The amount of money made over a period of time on an investment

For example, if an investment of $1000.00 yields a return of $100.00 in one year, then the return on the investment is 10%

Capital

The money and value of goods held by the company

Capital is used to acquire the products, equipment or services intended to help the company

Types of capital include

— cash from sales

— equity capital which is the value owned by individuals or organizations

— debt capital (borrowed funds)

Engineering Economics
Terminology of Engineering Economics

Depreciation

The decrease in value of equipment, buildings, etc., over time

— for example, a car is worth less after one year because of depreciation

— examples include
wear and deterioration

functional depreciation (inadequacy to perform the task necessary)

Benefit/Cost Ratio

The ratio of the worth of the benefits from and investment against the worth of the costs

— initial cost as compared to benefits, life of the investment, expected return, and cost to maintain the investment are all taken into account

Inflation

The phenomenon of rising prices which cause a reduction of purchasing power over time

— The same can of soda that sold for 25 cents one year and sells for 50 cents a few years later is due to inflation; inflation must be calculated for every investment

Page 3 of 6

Engineering Economics
Concept of Time Value of Money

Time Value of Money

The value of a dollar in the current market as compared to some future time

Inflation decreases the value of money

— example: fifty years ago, \$10 could buy a week's groceries and pay some of the rent; now it won't buy 2 movie tickets

Investments increase the value of money

— example: in 1626, Peter Minuit purchased Manhattan Island from the Native Americans for about \$24 in trinkets; if the Native Americans had sold the trinkets and invested the \$24 at 10% interest, today it would be worth \$48 quadrillion

Simple Interest Example

I=Interest earned; P=Principal; N=Number of interest periods; i=interest rate per period

$I=(P)(N)(i) ==> I=\$100 \times 3 \times .10 = \$30$

after three years the \$100 is worth \$130

Engineering Economics
Justifications for New Equipment

To Justify a New Product
- Sales projections (for figuring mfg. volume)
- Product definition (specifications)
- Results from pilot project
- Cost to scale to full production
- Definition of the model for full production

Safety and Regulatory Issues
- Government requirements and approvals
- Data from actual tests
- Standard practices

For Increased Production or Relief of Constraints (Bottlenecks)
- Justification for increased sales
- Impact on production system
- Calculations showing model throughput

To Justify Labor Savings
- Cost to hire or cost for overtime
- Long term impact and justification

Improvement of Quality
- Market justification
- Impact on production
- Impact on system flow
- Scrap and rework

Page 5 of 6

Engineering Economics Examples

Time value of money

What is the value of 1 cent if the amount is doubled each month for two years?

Year 1

J	F	M	A	M	J	J	A	S	0	N	D
2	4	8	16	32	64	128	256	512	1024	2048	4096

= 40.96 after one year

= $167,772.16 after two years.

Future value

What is the equivalent value of $5000.00 which is available now, if it were still available six years from now, assuming 12% interest?

$F=p(1+i)^n$ $F=(5000)(1+12)^6$ = $9869.00

Payback

A $40,000 machine is expected to be obsolete in 10 years with no salvage value. During its lifetime it should generate additional pretax profits of $8,000.00 per year. The additional taxes are $3,000.00. What is the payoff period (not time adjusted)?

Payback = (Initial - Salvage)/profits per year
= 40000-0 / 8000-3000
= 40000/5000

Payback = 8 years

Secton 9— Communication Skills

Drew Barnes — Reflections on Communication Skills

We found out that one of the most frequently mentioned areas of improvement needed by engineers is communication skills. The ability to communicate, both in writing and orally, will do as much (if not more) to maximize our contribution as engineers than any other skill. Developing good communication skills will also do more to ensure growth and promotion than nearly any other skill. The ability to clearly explain ideas, problems, solutions, and decisions in a way that others readily understand and does not offend, contributes greatly to the mission and goals of all organizations. This makes a lot of sense, but I thought being a good technical expert would be most important.

Technical competence is critical, but is almost assumed, at least from any respected educational institution. Therefore, the next most critical attributes are ethical behavior, good communication skills, and an ability to effectively solve all kinds of problems. In the class discussion the term "soft skills" was used a lot. "Soft" is a word used to describe skills or attributes that do not require heavy analytical expertise. These soft skills are often the ones that make the engineer successful over a longer period of time. The professor also talked about how the soft skills are interrelated. For example, to become a good writer one ought to be a good reader. And not just of technical material, but also all kinds of good literature. Being a good decision maker and problem solver will go hand in hand with good communication and good ethics. All of these attributes absolutely require the ability to think clearly, critically validate one's thinking, and express the result thoughtfully and with sound reason and judgment.

We were told that we would find numerous opportunities to use our communication skills. Could we get by at work without having to write or speak out? Probably, but just as a student who tries to get through school without being able to read well will eventually hit a roadblock, so will the engineer who attempts to get through work without developing and using good communication skills, regardless of their technical skills.

I was surprised to learn that some engineers estimate that 50 to 70 percent of their time is spent in some type of communication. The stereotype among many engineering students is that most of their time will be spent calculating, designing, or testing, when in reality it is the smaller part of their functions. And as one moves into management, the estimate of time spent communicating increases to as much as 90 or 95 percent. The ironic thing in all of this is that the better one becomes at communicating, the less time it will take to do it well. At least, the less time it will seem to take, as the better you get the more you will enjoy it. Being willing to develop and use good communication skills will open up many chances for individual growth and promotion.

I also learned that writing is a natural extension and indication of one's thinking abilities. As you write you are able to clarify in your mind what you understand and what you don't. As an example, we were given the assignment to write in a couple of paragraphs describing Newton's first law of motion (which states that a body at rest stays at rest and a body in motion stays in motion with a constant velocity until acted upon by a force) in terms simple enough that an 8-year-old could understand it. When I tried it, I found that I knew in my mind what the law meant, and I could even visualize it, but I had some trouble describing it. As I wrote about it I began to understand it better.

We also learned some specific things about the writing process and structure, both generally and for technical reports. These are in the notes.

Assignment: Rewrite the following memo to contain all pertinent information, but avoiding extraneous material. Hint: It should be easily done in one page containing less than half the lines.

SLOANE REPORT-Stage 1

September 9, 1957

To: H. B. Dawson

From: N. B. Sloane

Subject: Audit of Payroll Deduction Funds, Exville Plant

The audit last week of the records of those employees at the Exville plant who have deductions for U.S. Savings Bonds withheld from their wages or salaries indicated that the account in which these funds are held at the First National Bank-the account R. J. Wright, Cashier-was in balance with the individual account cards of the employees' deductions. However, there has been a mishandling of the funds which represents what might be considered collusion between Mr. R. W. Smith, the plant manager, and the cashier, although both of them advise that, while the procedure in which they have been engaging is admittedly not a very common one, each of the fails to see how his own actions in the matter can be considered in any way improper or otherwise deserving of censure. Smith has authorized payroll deductions for savings bonds since July 1, 1950, at which time Miss Platt informed the writer that it was decided by this company to undertake a concentrated buy-savings-bonds campaign for the purpose of convincing all of our employees to sign up to purchase bonds under the payroll deduction plan. Upon examination of Mr. Smith's deduction card it was indicated that although deductions from Smith's pay check have continued to be made since that time, Mr. Smith has not purchased any bonds. Smith is on a bimonthly payroll that is made up here at the main office. As you know, the total of all of the deductions on every payroll is forwarded to the cashier on every payday in order to be deposited by him in this account. It then becomes the duty of the cashier to purchase a bond for any individual employee whenever the individual accounts indicate that an employee has accrued a sufficient purchase price for one. As of the present time the plan is being taken advantage of by nearly 1,000 of our employees, 997 to be exact, therefore it has clearly achieved a considerable degree of success.

As was mentioned in the above paragraph, however, Smith has never purchased any bonds. The bimonthly payroll is written very early in the payroll period and as soon as this payroll is written the bond deductions which are listed on it are posted to ever individual employee's account. Therefore, very early in the month****Smith's bond deduction card has been credited with his****deduction. However, upon investigation of his individual deduction card it was ascertained that Smith has consistently requested of the cashier that the latter grant a refund to him of this $100 before the actual date of the payday for that period, the cashier always acceding to the request. This means that whenever the cashier has done this there was not really money in the bank for Mr. Smith's refund and therefore Smith, the plant manager, was actually obtaining advances in every single payroll period. The cashier, of course, would not have been able to acceded to the request if it had not been for the presence in the account R. J. Wright, Cashier, of funds that had been deducted on earlier payrolls from the pay of the other employees.

Feeling that this is not a proper procedure, it has been brought to the attention of the plant manager and also to the attention of the cashier. Mr. Wright, Cashier, says that he does not consider himself to have done anything for which anyone can justifiable censure him inasmuch as he was merely acting pursuant to the instructions to give Mr. Smith the refund that he has regularly received from the plant manager. Mr. Smith says that he does not feel that there have been any violations of the company's rules and/or policies and for this reason I have apprised him that I would include in my audit report. Mr. Wright will be 65 years of age on the 12th of October of this year and has been in the employ of this company since shortly after the First World War. In addition there is the further complication that the records indicate that Mr. J. J. Blake, the assistant plant manager, who was formerly at our Chicago, Ill., plant, while he has not utilized this procedure as frequently as Smith, on one or two occasions during the past several months, when for some reason or another he has been financially embarrassed, he has in effect likewise provided himself with a payroll advance through the same means. Mr. Blake, too, is on the bimonthly payroll that is made up here at the main office, whereas all of the other employees who have bond deductions withheld from their pay are on a weekly payroll that is made at the plant. There is not such doubt in my mind that you will agree with the two following conclusions:

1. That the actions of the above named personnel in this matter have not been within the intended results desired to be achieved by the "Buy U.S. Savings Bonds" campaign, and that
2. There is a loophole here by which it will be equally possible for any additional plant manager who might be so minded to perhaps also follow the Smith procedure, thereby also obtaining an improper advance on his salary prior to the day on which it actually becomes due.

Figure 8 This report is used by permission of Robert H. Burger and Associates.

Communications Skills
Overview

Importance of Learning Good Communication Skills

Types of Written and Oral Presentation Opportunities

Steps to Good Writing

General Structure of Written Material

Elements of a Good Paper

Expectations of College Writing

Plagiarism

Communications Skills
Importance of Communication Skills

Oral and Written Skills Are Part of an Engineer's Work

Good writing is particularly important

Use every chance to improve communication skills

Fifty to Seventy Percent of an Engineer's Time is Spent in Some Type of Communication

Managers Estimate That the Time Spent Communicating Increases to as Much as 90 or 95 Percent

The Better You Become at Communicating, the Less Time it Will Take to Do it Well, and the More You Will Enjoy it

Developing and Using Good Communication Skills Will Open Up Many Chances for Individual Growth and Promotion

Communications Skills
Types of Communication Opportunities

Examples of Written Opportunities

- Technical reports
- Memos
- Letters
- Proposals
- Product documentation
- Test results and explanations
- Meeting agendas
- Trip reports
- Performance appraisals

Examples of Oral Presentation Opportunities

- Product presentations
- Proposal presentations
- Sales presentations
- Conducting meetings
- Oral reviews of progress or design
- Normal daily interactions and conversations
- Organizational direction discussions
- Performance reviews
- Telephone conversations
- Negotiations
- Technical service calls

Page 3 of 6

Communications Skills
Steps To, and General Structure of, Good Writing

Three Common Process Steps of Writing Are
- Brainstorming and rough drafting
- Composing
- Revising and Rewriting

General Structure for Good Writing

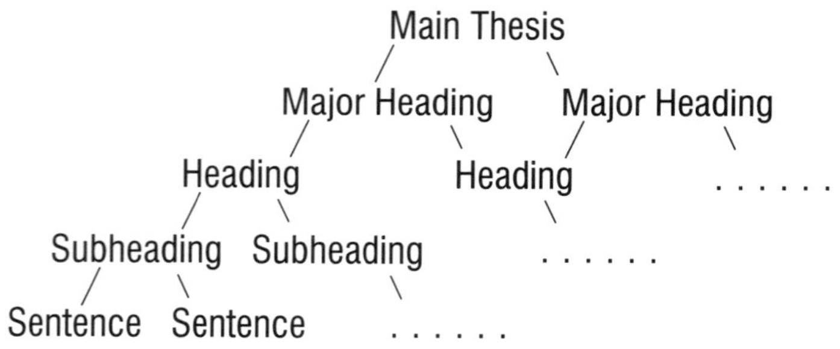

```
                              Main Thesis
                             /            \
                   Major Heading        Major Heading
                   /          \          /          \
            Heading             Heading               ......
            /     \                   \
   Subheading   Subheading           ......
    /      \             \
Sentence  Sentence      ......
```

The thesis gives direction and focus to the paper, sections, and paragraphs.

Communications Skills
Elements of a Good Paper

Invention

What to say
How to say it
What's the purpose in saying it (most important)

Arrangement

How you want the paper to play out will depend on audience and your purpose for the paper

Style

How you convey to your audience what you want to say
How you combine and use words and sentences
Use of grammar and punctuation

Delivery

Format
Neatness
Handwritten or typed
— handwritten papers are seldom acceptable today

*These four steps are adopted from *A Brief Guide to Writing Student Papers*, published by General Education and Honors College at Brigham Young University, 1990.

Communications Skills
Expectations of College Writing

High school writing tends to be evaluative writing, for the purpose of proving that a book was read or an experiment completed. College bound students often come to school believing that college writing will be the same.

College writing will be, we hope, more thoughtful and inquisitive. More important than "did you read the assignment" (that will be assumed and expected) will be

What did you get out of the assignment?
Do you agree with the message of the author and the author's way of presenting it?
Was the reasoning of the author clear and sound and how so?
What connections to experiences are there with other areas you are studying; or what did you learn from reading current issues in your field of study?

Plagiarism

Using another's work and representing it as your own
- — it does nothing for your learning
- — it can ruin your reputation and good name
- — it robs you of the growth and learning intended

Any student who intentionally plagiarizes should expect serious consequences, such as a failing grade on the assignment (at the very least) and perhaps in the class.

Page 6 of 6

Section 10 — Creativity and Innovation

Drew Barnes — Reflections on Creativity and Innovation

I always thought that creativity was the ability to daydream and come up with crazy ideas and abstract concepts. In a way it is, but I have learned now that it really is more than that. What was most helpful for me in this class was to find out that all of us are, or at least can be, creative. We were not just told this, however, we were shown how this can happen. The professor described it as being able to link knowledge from two separate subjects in a new way to solve a problem or fashion a new concept. Then the professor went on to explain how we can improve our ability to be creative.

We also learned that being truly creative, especially in engineering and technology, is greatly enhanced by improving our ability to think. Developing concentration and thinking skills will help develop the creative function.

There were a few examples given about creativity that were very interesting. One of them had to do with the number of problems on the failed Apollo 13 mission. They had an explosion on board and had to return to Earth without stopping at the moon. Numerous solutions to problems were found under extremely tight constraints and very limited resources in little time. The success of that mission was based on a few people being very creative and innovative in a very short period of time.

We talked about a few specific cases of creative thinking and watched some interesting short clips about creativity. Some important characteristics of creativity include being able to take risks and to accept and learn from mistakes. These logically go together. If I am going to take a risk, this inherently means I might fail, which means I made a mistake. But the mistake itself is a learning opportunity. Another characteristic that is very important is persistence. This makes sense as well. If you take risks and fail at some attempts, success will only come as we keep working. It seems to me that creativity and innovation are just plain hard work with the ability to think well and continue learning thrown in. So much for my abstract view of creativity.

We were given the example of Thomas Edison as a creative person. Very hard working, almost fanatically so, and made thousands of mistakes but kept going. When Thomas Edison's laboratory was virtually destroyed by fire in 1914, much of Edison's life's work went up in flames. At the height of the fire, Edison's 24-year-old son found Edison calmly watching the scene. "My heart ached for him," said his son. "He was 67—no longer a young man—and everything was going up in flames." But when Edison saw him he shouted, "Charles, where's your mother?" When I told him I didn't know, he said, "Find her. Bring her here. She will never see anything like this as long as she lives."

The next morning, Edison looked at the ruins and said, "There is great value in disaster. All our mistakes are burned up. Thank God we can start anew." Three weeks after the fire, Edison delivered his first phonograph.

We were given ways to work at building our creative ability and also taught how to be more confident in our abilities. A list of things that stop creativity was given to us as well. I looked and could sure see how I have used them to stop my own thinking.

Another idea that was explained was the benefit of constraints. I guess I had always thought that constraints were bad things, that they hurt the creative process. I have learned that constraints are what helps the creative process work. It is clear that we need to understand what are real constraints and what are imposed by our own thinking, but the whole idea that "necessity is the mother of invention" is the idea that constraints promote invention.

Assignment: Complete the following problems.

1. Your house is on a corner lot, and you have a beautiful lawn. There is only one problem. The neighborhood kids keep cutting across your lawn, wearing a very visible path across it (just like what happens on most college campuses). List fifty separate ways you could use to keep kids from cutting across the lawn. Constraint: no killing or maiming allowed.
2. Find and describe an innovative product, process, or idea. Explain why you find it to be so creative or innovative. Could you improve on it? If so, how?

Creativity and Innovation
Overview

Improving the Effectiveness of Creativity

Developing Good Mental Skills

Working at Being Creative

Building Creativity Confidence

Improving Perceptiveness

Move Out of a Normal Problem Environment

Impediments to the Creative Process

Page 1 of 5

Creativity and Innovation
Improving the Effectiveness of Creativity

Learning to Search Your Mind

- Develop the ability to look at all subjects, even seemingly unrelated ones
- Identify and establish links between subjects for the current problem
- Use the links to help encourage solutions for the problem
- Acquire technical and general information in many areas

Developing Creative Thinking Skills

- Improves the ability to think "out of the logical box"
- Builds concentration abilities
- Builds knowledge base for increased mental capacity
- Helps teach the ability to think critically and practically

Page 2 of 5

Creativity and Innovation
Improving the Effectiveness of Creativity

Working at Being Creative

Creativity is a process, not a product

Be persistent

Do something; action usually helps creative thoughts

Become emotional and enthusiastic about creativity

Building Creative Confidence

Assert and champion a new concept

Take a risk

Be prepared for criticism

Seek help from weaknesses

Understand that everyone is capable of being creative

Creativity and Innovation
Improving the Effectiveness of Creativity

Improving Perceptiveness

- Rely on intuition, imagination, and impetuousness
- Envision the consequence
- Recognize, accept, and learn from mistakes
- Slow down and look at the big picture

Move Out of a Normal Problem Solving Environment

- Use role plays
- Develop and use analogies
- Rid yourself of normal impediments to innovation
- Use free association methods, brainstorming
- Ask questions like, "What if I was not afraid of failing?"
- Turn on music, go for a walk, etc.

Page 4 of 5

Creativity and Innovation
Improving the Effectiveness of Creativity

Impediments to the Creative Process

Believing there is only one “right answer”

Creativity is taught out of us; seek to get it back

Not being able to suspend “logic”

Not believing you can learn to be creative

Not willing to reject a “bad” rule

Not asking “What if?,” “How?,” and “Why not?”

Believing mistakes are always detrimental

Not believing a solution exists

Thinking that play is frivolous

Avoiding ambiguity

Believing you are not creative

Section 11 — Ethics and Professionalism

Drew Barnes — Reflections on Ethics and Professionalism

The lecture on ethics was another eye-opener for me. Not that I am a dishonest person really, but it made me look at what I think and what I do more seriously. I guess a lot of what I am finding in these lectures about subjects and skills that are not purely technical, the "soft" skills, is that engineers and technologists are well respected as a group. For example, we learned today that studies and surveys have shown that engineering is considered by many to be among the most ethical of professions. That makes me feel good about what I am studying, but at the same time, as the professor said, it also puts some pressure on me to live up to that standard. Not that I expect to have problems, but I don't want to shame my profession.

A lot of what we talked about today is pretty straightforward and common sense. One aspect that was raised in class by the professor was from a paper written by some guy back in the late 1960's. It was kind of contrary to the view I have of the 60's, that it was the "me" generation. The essence of the paper was that the problem, in terms of what is ethical and right, isn't that people break laws and rules. There have always been people who have done that. The problem is that too many people aren't willing to acknowledge that there is a set of established principles by which to live. Throughout history people have subscribed to principles of right behavior by looking to either philosophy or religion for guiding principles. The task was to understand and live by these higher order principles. The point was that although philosophy and religion didn't necessarily agree, they both knew that the principles were defined above themselves. Now, according to the paper, too many people see the principles as changing based on the situation (what the professor called situational ethics) or that they did not even believe there was a guiding set of principles. Therefore, whatever they chose to do was OK. This leads to anarchy and a society without order.

The professor then described how this idea also affects our professional behavior. Professional behavior is the attributes we adhere to as part of the professional education and training we have and as part of what is expected of us as educated people.

All of this was very interesting and made me think about what I see as guiding principles and fundamental concepts. We were also given a process for working through ethical problems. I have one example in my notes. We talked about some questions to ask ourselves when we are faced with ethical questions that help us make the right decision. The first question is "Is it legal?" If it isn't, we had better stop right there. But that is just the first question. The second one is "Is it balanced?" This means, does it give unfair advantage over someone else? If so, we had better consider if it is really right or not. Then the last question is a good one to finish with. It is, "How would it make me feel about myself?" This one should make us stop and think hard about what we do. Along with it, someone brought up the idea that we should consider how we would feel if people we care about, like our family or close friends, found out about our choice. Would they be proud of me or not? And more important, would the choice be one we would want them to find out about? Good questions to help me make good choices.

The material on professionalism is basically how we ought to act and treat people in important areas but are not really considered ethics. A lot of them were habit types of things in terms of neatness, promptness, and such. I do need to make some improvement in these areas, and the idea that doing these things well will make me a better engineer and a better person help provide motivation.

The Flappy Lips at the Concert

Case Study

You and a friend have just gotten to your seats at a concert you have been looking forward to for some time. Within a minute or two you realize that two men behind you are evidently in a business similar to yours. They are your competitors and are talking about a product that is proprietary. In fact, your company has a product that is similar, and you are on the product team. You also know, through the industry grapevine, that this competitor is about 6 months or so ahead of you in projected product release.

You have a good 20 minutes before the concert will start, and your friend is talking to another friend who is sitting nearby.

What do you do?

- Sit quietly and take notes.
- Ask them questions to get more information.
- Leave the concert.
- Wait in the foyer out of earshot until the concert starts.
- Tell them who you are and suggest they might want to stop talking about the product.
- Ask your friend to help you take notes in case you miss something.

Write a paper describing the option you would take (it may be different from those listed) and why. Include a brief discussion on the correctness of each of the options listed and a justification of the option you chose.

Ethics and Professionalism
Overview

Engineers and Ethics

Background and rationale

Fundamentals of a code of ethics

A process for solving ethical dilemmas

Professional Expectations and Behavior

What a professional is

Characteristics of professional behavior

Advantages of professional behavior

Ethics and Professionalism
Engineers and Ethics

Background and Rationale

Ethics is not just how you think, but how you act

Must be fundamentals based

Correct behavior is the basis for human success

Poor ethics cannot be compensated for by any other strength

Civilizations are all about how people are treated

US constitution is founded on individual responsibility

Use of knowledge should be based on virtue and values

"Knowledge derives its value from this, that it enlarges our power, and directs us in the application of it. For in the right employment of our active power consists all the honor, dignity and worth of man; and, in the abuse and perversion of it all vice, corruption, and depravity.

And every man must acknowledge, that to act properly, is more valuable than to think justly or reason acutely."

Thomas Reid, 1813

Ethics and Professionalism
National Society of Professional Engineers—Code of Ethics

Preamble

Engineering is an important and learned profession. As members of this profession, engineers are expected to exhibit the highest standards of honesty and integrity. Engineering has a direct and vital impact on the quality of life for all people. Accordingly, the services provided by engineers require honesty, impartiality, fairness and equity, and must be dedicated to the protection of the public health, safety and welfare. Engineers must perform under a standard of professional behavior which requires adherence to the highest principles of ethical conduct.

I. Fundamental Canons

— Statements regarding competence, acting in a truthful manner avoiding deceptive acts and conduct that is honorable, responsible, ethical, and lawful.

II. Rules of Practice

— Engineers' responsibilities with respect to welfare of the public, performing services only in the areas of their competence, issuing public statements, etc.

III. Professional Obligations

— Statements that engineers shall be guided in all their relations by the highest standards of honesty and integrity.

Page 3 of 5

Ethics and Professionalism
A Process for Solving Ethical Dilemmas

Define the Problem or Dilemma as Much as Possible

Identify Any Assumptions That Need to Be Made

Identify the Fundamental Laws, Principles, and Guiding Values That Appear to Be Operating in the Dilemma

If the Problem Is Complex, Break It into Solvable Parts

Apply an algorithm for the solution of the dilemma

- — is it legal? (does it adhere to federal, state, and local laws?)
- — is it professional? (does it adhere to your technical society's code of ethics?)
- — is it moral? (does it adhere to religious/moral position?)
- — are there any conflicts that can be eliminated? (or at least minimized?)

Make Your Decision

Seek for Confirmation of the Decision

Act Out the Decision, Assess the Consequences, and Seek Continued Improvements to Your Solution

Ethics and Professionalism
Professional Expectations and Behavior

What a Professional Is

Acting in accordance with the integrity and prestige of your profession
Engineering, law, medicine, business, etc., are all professions
A professional has a more fundamental knowledge of a discipline
A professional also has a broader perspective
A professional treats others respectfully

Characteristics of Professional Behavior

A professional maintains responsible control of work habits and environment

Attributes that demonstrate this attitude are

— promptness
— neatness
— preparedness
— appropriate appearance
— safety
— proper and considerate speech
— integrity
— honesty

Advantages of Professional Behavior

Reputation
Trust
Will be seen as able to represent groups and organizations
Treated well by others

Page 5 of 5

Section 12 — Company Organization and Corporate Economics

Drew Barnes—Reflections on Company Organizations and Corporate Economics

This was another business type of class. We learned about how to look at a profit and loss statement to see how a business was spending its money and a few terms regarding business balance sheets. I know I will need to understand this better someday, but for now what we got seems like enough.

The first thing we talked about, however, was the organizational structure in companies. Three different structures were described. The hierarchical, the matrix, and the flat organizations were explained and some characteristics of each of them were given. I was trying to think about ACTC and which one they use. They have a president, and vice presidents and directors and such, so that seems like the hierarchical. But they also use a lot of teams and that is an attribute of flat organizations. They also have product groups as well and the functional groups, and we learned that product groups are attributes of the matrix organization. I decided that maybe combinations of these types are used and that the overall form of ACTC is hierarchical, but that it tries to also take advantage of the other types in some situations. Maybe a lot of companies are like that. One thing that seems to be fairly certain is that teams are used by more and more companies. According to the professor, some use them well, and some don't.

After the discussion on organizations, we talked about financial aspects of companies. We learned some terms and learned how to read balance sheets and profit and loss statements as well. In a way, it really is just common sense. Basically, it is similar to my own financial situation. I make money working at ACTC. I have expenses that I have to pay. If my expenses are less than what I make, then I am profitable; if not, then I have a loss. It isn't really that simple because of all the other things a business has to worry about with equipment, inventory, sales, and investments. But as far as what we call the bottom line, that's how it is.

One thing that was most interesting to me was the fixed and variable costs. In my own life I have the same thing. My fixed costs are dorm fees, tuition, and the monthly amount I pay to eat at the cafeteria. Then my variable costs are the ones like entertainment, and junk food, and even some of my school supplies, etc. If I am below my breakeven point, which is the point where my income at least is the same as my expenditures, then I have to increase my income or reduce my expenditures. The fixed costs usually can't budge much (that's why they are called fixed), so I have to do something with my variable costs. But that would be much easier for me to do than a company. This is because part of a company's variable costs include material for building a product. If you don't build a product then you don't sell anything, and therefore you don't have any revenue or income. That is a variable cost that you have to be very careful about reducing. So we learned that in a company, you need to learn to manage all your costs as effectively as possible.

This is important for us in engineering and technology to understand because we are involved in the areas of the company that spend a lot of money. Therefore, beyond just being able to see how our company is doing financially, we also play a role in how effectively the company is able to manage the resources it has. If we are in a company that produces things and are able to find a way to reduce some of the variable costs by 10 percent, then the effect on the bottom line of the company is very good. This would make us more competitive and a far healthier company.

Assignment: Obtain a copy of a company financial report, review it, and determine if the company is healthy enough to invest $10,000 of one's own money. The report should include a brief description of assets, liabilities, balance sheet information, some idea of the fixed and variable costs in the company, and profits. This report will likely require two or three pages to complete.

Company Organizations and Corporate Economics Overview

Organizational Structures

- Purpose of organizational structures
- Hierarchical
- Matrix
- Flat

Business Finance

- Profit and loss statements
- Definition of terms

Balance Sheet

- Assets
- Liabilities
- Stock
- Equity

Company Organizations and Corporate Economics
Organizational Structures

Purpose of Organizational Structures

- Vehicle for management of the organization
- Defines the relationships between functions
- Provides the structure for organizational planning
- Defines the framework for dividing resources
- Describes the outline for staffing needs and control procedures

Hierarchical or Functional

- Top down controls
- Authority lines well defined
- Typically has many layers of management
- Usually highly bureaucratic in nature

Matrix

- Combines functional responsibilities with project responsibilities
- Reporting lines sometimes get confused
- Can save resources over pure functional or hierarchical organization through shared costs
- Higher quality of service due to focus on expertise

Flat

- Usually team based
- Authority and responsibility is spread out
- Control and decisions pushed to the lowest level
- Encourages employee empowerment but less efficient

Company Organizations and Corporate Economics
Business Finance

Profit and Loss Statements

- Available only for public companies (usually)
- Shows revenue, sales, and other income
- Describes cost of goods sold
    - — direct costs—based on amount of product made
    - — indirect costs—support of product
    - — fixed costs—support of plant and facility
- Shows the gross margin
- Total expenses are described
- Gives depreciation amounts
- Net income or the bottom line

Definition of Terms

- Fixed costs are
    - — rent
    - — insurance
    - — property taxes
- Variable costs are
    - — material
    - — labor
    - — electricity
    - — sales expense

Company Organizations and Corporate Economics
Balance Sheet

Shows the Wealth and Financial Position of the Organization

Outlines the owned, expected (both assets), and owed (liabilities) resources by the company

Assets

- Cash from sales, etc.
- Accounts receivable
- Equipment and supplies
- Most inventory (in terms of product owned)
- Cost of inventory can also be considered a liability
- Investments
- Buildings

Liabilities

- Accounts payable
- Mortgages
- Outstanding bonds or debts

Equity

- Stocks, preferred or common
- Retained income (historical health of the company demonstrated by the flow of past transactions—not cash)

Page 4 of 4

Section 13 — Social and Environmental Issues

Drew Barnes — Reflections on Social and Environmental Issues

A thought provoking class on social and environmental problems and how we in the fields of engineering and technology ought to be aware of and help solve them. The professor explained that technical professionals are often unfairly stereotyped as ignorant and uncaring about social and environmental issues. But in fact, we can help solve many of these problems because we are oriented toward problem solving. We simply need to have the interest and knowledge of areas outside our technical expertise.

The tremendous rate of technological advancement and automation has created a need for technical professionals who can use the latest technology effectively, ethically, and cooperatively, and who are adept in communication, building effective relationships, possess a strong set of moral values, and are prepared to be more mobile in terms of function and professional responsibilities. They need knowledge on a broader range of subjects.

From an educational standpoint, it is impossible to keep abreast of such rapid and diverse technical progress. With changing social and cultural conditions, understanding more than just specialized subjects is crucial to one's long-term success and well-being. The professor quoted Paul E. Gray, former president of M.I.T., who said, "The greatest professional challenge our students face will not be purely technical. Rather, they will be interwoven with economic, social, and ethical considerations. To act responsibly and professionally, our graduates must have not only the ability but the inclination to view problems, their possible solutions, and their consequence in a manner that draws on and ties together various domains of knowledge."

The professor also quoted Albert Einstein who said "The significant problems we face cannot be solved at the same level of thinking we were at when we created them." We were told that, at least with respect to social and environmental issues, the method and mechanism for solving problems on a higher level is the simultaneous application of technical skills and expertise with mind broadening effect of a general education influence. Because of this closely dependent relationship, and the nature of engineering and technology to create, innovate, and develop solutions, engineering education is a very appropriate practice field for a greater appreciation of general and liberal education concepts.

The lecture taught us how we could become more aware of social and environmental concerns. We were also encouraged to closely examine the policies and procedures of our companies, and see how the proper balance between company profit and social and environmental concerns could be reached. The last point we talked about is the need to view a product's design and production throughout its life cycle including disposal and/or recycling. I thought this was a good class because there were many concepts that I hadn't considered before. Shibusawa Eiichi, who Frank Gibney called the conscience of Japanese Business, was a student of Confucian philosophy. He said, "Morality and economy were meant to walk hand in hand. But as humanity has been prone to seek gain, . . . scholars misunderstood [Confucius's] true idea. . . They forgot that productivity is a way of practicing virtue."

Social and Environmental Issues Overview

Developing an awareness of long term consequences of social and environmental concerns

Be able to balance social and environmental concerns with productive corporate policy

Consider social and environmental consequences in routine corporate decisions

Consider the impact of a product's full life and disposal on social and environmental issues

Gandhi's social ills

Social and Environmental Issues Awareness and Balance

Develop Awareness of Consequences of Social/Environmental Concerns by

Working to learn more about the social concerns or your time and area

Studying material that is unbiased and directed to wise care of the environment

Taking advantage of classes and subjects outside your major and area of technical expertise

Avoid extreme, illegal, and unethical positions, groups, and behavior that claim to be concerned about social or environmental issues

Balance Social and Environmental Concerns with Productive Corporate Policy by

Consulting with trusted sources outside the organization about the impact of the organizations on social and environmental items

Understand the corporate position on social and environmental points

Strive for objectiveness regarding the decisions and policies put in place by the organization

Page 2 of 4

Social and Environmental Issues
Consequences of Corporate Decisions and Product Life

Consider Social and Environmental Consequences in Routine Corporate Decisions by

Striving to understand the impact of layoffs and other organization restructuring

Striving to ensure safety concerns are always reviewed and followed

Be a good citizen

Consider the Impact of Product's Full Life and Disposal with Respect to Social and Environmental Issues

Evaluate the impact of product production and disposal during the design stage

Strive for recycling options with production waste and obsolete products

Creatively search for profitable solutions to product waste and disposal

Page 3 of 4

Social and Environmental Issues
Social Ills

M. K. Gandhi's Seven Social Ills

Politics without principle

Pleasure without conscience

Knowledge without effort

Wealth without work

Business without work

Science without humanity

Worship without sacrifice

Section 14 — Lifelong Learning

Drew Barnes—Reflections on Lifelong Learning

This class, like to the one on social and environmental responsibility, addressed issues not normally considered technical. Some of the points, however, were directed towards how to stay abreast of state-of-the-art technology in our field.

The professor said that the tremendous rate of technological advancement and automation has required technical professionals to learn at a faster rate than ever before. This class focused on how to do that and also encouraged us to learn about other areas besides technical. The professor said it is part of being a good engineer and a good citizen.

We talked about what the professor called 'not letting our education get in the way of our learning.' Learning is what I am at the university for. It's the idea that one learns how to think, innovate, evaluate, and draw conclusions based on a set of information, principles, and values tried by history and proven by wisdom.

We also talked about a number of things that can help us learn throughout our lives. Professional organizations are good sources. They all offer regular meetings, publications, videos, and seminars targeted at helping members learn new skills and technology. They also offer us the opportunity to meet others in the field. The professor told of how he has been able to get help solving problems by contacting people he met at professional meetings. He also told a story of how, when he was in industry, he and his company were able to help another engineer and his company through a tough time, and how they knew each other from professional society meetings.

We also learned that many companies have their own educational departments and that others try to provide means for further education. Those with departments might offer classes on subjects important to the company. These classes help solve company problems and help employees learn more about the company's product, services and needs. The professor told about a friend of his who said taking in-house courses not only helped him learn about the company, but also helped get him promoted. What I didn't know is that many companies will pay the yearly membership fee for professional organizations and will sometimes pay for attendance at professional conferences and seminars. This encourages their engineers to belong to professional organizations.

We were also encouraged to have hobbies and outside interests. There are benefits in learning, and it's good to have an outlet in times of stress.

We finished class with a quote by James Barcus, former president of the Society of Manufacturing Engineers. Regarding the work of a special committee for Lifelong Learning and Career Development, he said, "The committee believes lifelong learning is emerging as the most important competitive consideration." And then, ". . . the need for work-life quality that ensures maximum productivity takes on new meaning—and so does learning. In fact, learning how to learn may become our number one priority."

Lifelong Learning
Overview

Purpose of College
- Learning to learn
- Teaches discipline
- Commencement to a life of contribution and growth

Professional Organizations
- Purpose of professional organizations
- Networking with peers

Taking Advantage of Company Resources
- Educational opportunities
- Seminars

Hobbies and Interests

Preparing for Changes in Career and Life

Lifelong Learning
Purpose of College

To Learn to Learn

- The learning process will be necessary throughout life
- Much greater satisfaction from career and other areas of life
- Requires breadth and depth for best lifelong learning

Why Cheating is Detrimental to You

- Stops the learning process
- Hurts reputation
- Impedes the discipline required for learning

Why General Education Classes

- Promote a broader view of life and the world
- Introduce concepts from outside disciplines that will promote thinking ability
- Broader interest and knowledge promote most interesting life

College is the commencement, not the conclusion, of learning, contribution, service, and growth.

Page 2 of 4

Lifelong Learning
Professional Organizations

Purpose of Professional Organizations
- Keep abreast of new developments in the field
- Provide seminars and classes
- Strengthen educational programs in the field
- Send a united voice regarding issues related to the field
- Give opportunity for leadership and personal development

Peer Learning
- Fosters friendships and contacts that will help throughout life
- Encourages creativity as new perspectives are gained
- Provides opportunity to present new ideas

Lends Credibility to New Advancements
- Testing and validation of new innovations
- Sounding board for new ideas
- Board and codes of ethics to ensure appropriate application of new innovations

Staying Current on Technology
- Critical to stay current on technology to stay employable
- Short courses, seminars, personal research
- Also helpful are periodicals, journals, etc.
- “Half life” of current technical information approximately 4-8 years

Lifelong Learning
Taking Advantage of Company Resources

Company Educational Opportunities

- Companies encourage education in many different disciplines and areas of expertise
- Many companies offer in-house training on technology and skills related to your work
- Opportunities for graduate work paid by the company
- Companies will also often pay for membership in professional organizations
- Companies will also often pay for attendance at professional society meetings or other professional seminars

Hobbies and Interests

- Keep life interesting and fun
- Provide stress release
- Good for family relationships
- Encourages learning

An attitude of lifelong learning helps prepare one for changes in career and life.

Section 15 — Safety

Drew Barnes—Reflections on Safety

Though this was discussed briefly in class at school, I learned more at work about the real practice of safety. Because ACTC is a manufacturing plant and has a lot of equipment and situations that could potentially be very dangerous, it really focuses on safe operations. It is, in fact, the first priority. I hadn't previously considered how important it is. Of course, if someone is hurt, it not only costs the company in terms of medical costs and increased insurance, but it also costs the company the time lost from that employee. Poor safety also affects employees' morale. These costs add up fast and can be very significant.

It was explained to me that safety isn't just what they do at work. Safety is a mind set or attitude in how people live. For example, I learned that presentations have been given at ACTC on careful driving, safety at home, and safety in leisure activities. If an employee has an attitude of safety in how they drive, how they play, and in their home, then they will also be safe in their work.

The notes I have show some basic safety rules and an example of a safety inspection sheet used at ACTC. More specific rules are developed at each area of the plant by those who work there. In fact ACTC has an interesting way of teaching and checking safety. A team of three or four people made up of employees from various areas of the plant comprise the safety team. This team assignment rotates among all the employees, each taking their turn every year or two. This team does a safety inspection each month and notes all violations they see. Then they meet with the other employees to discuss this inspection and any violations to the rules they have noted. The meeting covers any changes that need to be made in order to increase the safe operation of the plant. In this meeting, one of the team members also gives a presentation on safety. The presentation can be regarding work, home, or other areas, just as long as principles of safety are taught.

I think this method is a very good one. It gets everyone involved and encourages ownership of safety in all employees. ACTC's safety record is very good. The company's safety record is much better than industry average and their safety rules and procedures are much more applicable than those required by OSHA (Occupational Safety and Health Administration). OSHA is a government organization that focuses on safety, but can't do near the job that people in the plant can do.

I learned some good things about safety and the importance of being safe, neat and organized. I even found some safety issues in the room I share with Stu. For example, after learning about it at work, I came home and realized that we have a power strip with too much stuff plugged into it. This could be a hazard, especially when it is laying on the floor and we start piling dirty clothes and books on top of it. The principles of safety are things that will not only help me be more safe, but also better organized and neat. But according to what I learned, having those qualities will also help me be more safe.

Safety
Overview

Safety Rules

Safety glasses (with side shields) in all areas with moving equipment and materials

No loose clothing or inappropriate clothing
- — ties tucked in or removed
- — shirts buttoned
- — no long belts or other hanging items
- — no shorts

Appropriate footwear
- — no sandals
- — no open-toed shoes
- — steel-toe shoes preferred (many companies offer discounts on shoes through the safety office)

No running or foolhardy play in the plant

Safety rules will be specific to the plant or area in the plant. These are general rules that commonly apply to workplace safety.

Page 1 of 2

Safety Inspection Sheet
(sample)

Plant Area: ______________________________

1. Are safety rules being followed:
   Comments:

2. Do hazardous conditions exist:
   (chemical storage, oil on the floor etc.)
   Comments:

3. Are tools in their proper place:
   Comments:

4. Is housekeeping good:
   (floors clean, drawers closed, is area generally neat and tidy)
   Comments:

5. Is work in process material properly stored/located?
   Comments:

6. Other factors which might be unsafe:
   Comments:

Page 2 of 2

Index of Subjects by Chapter

Chapter One
Mind teaser puzzle

Chapter Two
Mind teaser puzzle solution
Networking for a job search
Professionalism, characteristics of

Chapter Three
Professionalism, characteristics of
Definition of engineering and technology
Spectrum of science, engineering, and technology
Practice of engineering & technology
Mathematics as the language of science, engineering, and technology

Chapter Four
Goal setting methods and principles
Priorities
Academic success principles
Personal success principles
Study skills

Chapter Five
Professionalism, characteristics of
Courage in looking for a job
Networking in a job search
Safety in the workforce

Chapter Six
Characteristics of engineering and technology
Communication skills
Relationships with others
Concepts of calculus

Chapter Seven
Need for communication between engineering functions
Technical part description
Practice of engineering and technology
Deciding on a major
Studying skills
Impact of engineering and technology on history
Trigger inventions
Society problems solved through technology

Chapter Eight
Data, importance of
Pareto chart
Eighty-twenty rule
Scientific method
Practice of engineering

Chapter Nine
Practice of engineering
Benefits of a career in engineering and technology
Role of engineering profession in society
Problem solving in the basic sciences (physics)

Chapter Ten
Listening
Communication
Discipline & persistence
Study skills
Data analysis & charting
Team work

Chapter Eleven
Correlation chart/scatter plot
Scientific method
Problem solving
Practical application of problem solving (car probelm)

Chapter Twelve
Problem solving
Design/work log notebook
Data analysis
Practice of engineering

Chapter Thirteen
Social and environmental responsibilities
General education
Lifelong learning
Gantt chart
Project management
Concurrent/simultaneous engineering

Chapter Fourteen
Correlation chart
Annealing
Potential energy
Heat treating
Communications and group interaction
Education, importance of

Chapter Fifteen
Group study
Persistence
Engineering economics
Risk and return
Return on investment
Professionalism, characteristics of
Design/work log book
Characteristics of good experiments
Testing

Chapter Sixteen
Histogram
Data analysis and validation
Charting methods
Problem solving
Documentation of results
Annealing

Chapter Seventeen
Writing reports
Writing skills
Memo forms

Chapter Eighteen
Practice of engineering
Personnel development skills and
characteristics
Lifelong learning
Family

Alphabetic Index of Subjects with chapter reference

Topic	Chapter
Academic success principles	4
Annealing	14, 16
Benefits of a career in engineering and technology	9
Car starting problem	11
Characteristics of good experiments	15
Charting methods	10, 11, 13, 14, 16
Communications	6, 7, 10, 14
Concurrent Engineering	13
Correlation chart/scatter plot	11, 14
Courage in looking for a job	2, 5
Data analysis	10, 11, 16
Data analysis and validation	16
Data, importance of	8, 16
Deciding on a major	7
Definition of engineering and technology	3
Description of engineering disciplines	3, 9
Design/work log book	12, 15
Discipline	4, 10, 18
Documentation of results	16
Education - importance of	13
Eighty-twenty rule	8, 11
Engineering economics	15
Family	6, 18
Gantt chart	13
General education	13
Goal setting methods and principles	3
Heat treating	14
Histogram	16
Impact of engineering and technology on history	7
Lifelong learning	13, 18
Listening	10, 14
Mathematics as the language of science	3
Memo format	17
Need for communication between engineering functions	7, 13, 14, 16
Networking in a job search	2, 5
Pareto chart	8, 11
Personnel development	3, 18
Physics problem	9
Potential energy	13
Practice of engineering	3, 6, 7, 8, 13, 15 18
Practical application of problem solving method	11

Priorities 3, 7
Problem solving 7, 12, 15
Problem solving in the basic sciences (physics) 11
Professionalism 2, 5, 10, 18
Project management 13
Return on investment 15
Risk and return 15
Role of engineering profession in society 3, 7, 9

Scientific method 8, 11
Simultaneous engineering 13
Social and environmental responsibilities 13
Society problems solved through technology 7
Spectrum of science, engineering and technology 3
Study habits/skills 3, 6, 10

Technical part description 7
Testing 15
Team work 10
Trigger inventions 7

Writing skills 17
Writing reports 17

Index

80–20 rule, 63

advisors, working with, 175
algebra, 210, 215
annealing, 116, 140
Arab, innovations, 163
attitude, positive, 182

Bacon, Francis, 164
balance sheet, 250
Barcus, James, 256
behavior
  classroom, 10–11
  professional, 244
Ben Franklin method, 177, 183
benefit/cost ratio, 218
Boyle, Robert, 164
Burke, James, 57–58, 197
business finance, 249

calculus, 17, 47, 205, 212
capital, financial, 217
career
  choice, 44–46, 143–44
  in engineering, 55–56, 70, 71–73, 119–20, 191–92
    benefits of, 71–72, 73–74, 191–92, 197
  in technical field, 55
character development, 122
characteristics, personal, 154, 183
chart, correlation, 114–15
charts
  histogram, 136
  Pareto, 62–63, 66, 67, 83–84, 97, 98, 139
  scatter plot, 136–37, 138
Chinese, innovations, 163
classroom behavior, 10–11
code of ethics, 242
cognition, 184, 187
cold working, 116–17
college
  purpose of, 258
  writing, 230
communication, 17
  skills, 149–50, 222–24
  types of, 227
  written, 146–49, 228–30
company
  assets, 250
  costs, 245–46, 249
  equity, 250
  liabilities, 250
  organizations, 250
concurrent
  engineering, 108–09
  schedule, 107–08
Confucius, 251
constraints, 190, 232
  value of, 184
corporate economics, 245–50
correlation chart, 114–15
creativity, 190, 231–32, 233–37
  developing, 234
  impediments to, 237
  improving, 234, 237
customer, 190

da Vinci, Leonardo, 164
daily schedule, 170
data, 61, 91
  charting, 87–89, 97–98
  importance of, 60–61
  presenting, 61–63, 114–15
  trends, 90–91
depreciation, 218
design
  log, 128
  problem, 188
  process, 184, 189
designed experiment method, 141
discipline, 24, 32–33, 177, 183
documentation, 100, 101, 102
Doyle, 60

economics
  corporate, 245–46, 249, 250, 252
  engineering, 214–15, 216, 217, 218, 219, 220, 221
Edison, Thomas, 100, 196, 231
education
  continuing, 256, 260
  value of, 105–06, 120, 197
Egyptians, 161
Eiichi, Shibusawa, 251
Einstein, Albert, 251
energy, internal, 116–17
engineer, 163
  and society, 71
  salary of, 19
  work of an, 69–70
engineering, 14
  benefits of, 191
  as a career, 74, 119, 196, 197
  careers in, 59, 197
  concurrent, 108–09
  definition, 11–14, 194, 195
  development of, 65
  developments, history of, 161, 162, 163, 164, 165
  economics, 122–23, 126, 214–15, 216, 217, 218, 219, 220, 221, 224, 225–26,
  ethics, 238–39, 240, 241, 242, 243, 244
  fields, 151–52
  focus of, 14
  functions, 146, 195
  history, 58, 59, 157, 160–67,171
  impact on civilizations, 57
  and mathematics, 15

practice of, 191–92, 194, 195, 196
and science, 195
study of, 18
technology, 14, 159
definition, 18
focus of, 14
function of, 194
environmental
issues, 251, 252, 253, 254
ethical problems, 243
ethics, 238–39, 241, 242, 243, 244
code of, 242
professional, 244
experiments, conducting, 131

finance, business, 249
fixed costs, 245–46, 249
flat structure, 250, 253

Galileo, 164
Gandhi, 255
Gantt chart, 107–08
general education, 104–05
geometry, 205, 211
goals
achieving, 27, 177
achievement process, 30–31, 32, 33
commitment to, 28
following through on, 122
guidelines, 180
major, 25–26
measuring, 28
integrity of, 28–29
reviewing, 31–32
rewards for accomplishing, 30
setting, 25, 26, 27, 31, 32, 33, 35–36, 177–78, 180, 183
sharing, 30
writing, 29
Gray, Paul E., 251
guidelines, 131
for setting goals, 179–80
using, 128

habits
breaking bad, 153–54
developing good, 24, 145–46

heat treating, 116–17
hierarchical structure, 245, 248
histogram chart, 136
historical developments, 161–65, 166, 167
history
of engineering, 157, 161–65, 166, 167
of engineering developments, 161–65
honesty, 127
Hooke, Robert, 164
Hoover, Herbert, 69, 70, 86
hypothesis, 64–65, 66

Industrial Revolution, 165
inflation, 218, 219
information gathering, 90
ingenitor, 163
ingenuity, 157
innovation, 231–32
innovations, engineering, 157
integrity, 177, 183
internal energy, 116, 117
inventions, 157–58, 167
history of, 69
inventors, 164

job
interviewing, 41–42
satisfaction, 73
search, 38–39, 41

key inventions, 157–58, 166, 167

lab workbook, 196
Lakein, Alan, 25
Lavoisier, Anton, 164
lectures, 174
lifelong learning, 154, 196, 256, 257, 258, 259, 260

mass production system, 165
mathematical limits, 47–48
mathematics, 16, 17
in engineering, 15, 195, 205–06, 210, 211, 212, 213
matrix structure, 245, 248
memo writing, 146–49
Mesopotamians, 161
Middle Ages, 163

mindteasers, 8–9
Modern Age, 165
money, time value of, 214, 216, 219, 221

National Society of Professional Engineers, 242
networking, 7
Newcomen, Thomas, 165
Newton, Sir Isaac,157, 164
notebook, keeping a, 98, 99
notes, taking, 168, 174

oral
communication skills, 149–50
presentations, 149
organization, 171, 172
organizational structures, 245–46, 248
organizations, 248
business, 245–46
company, 248
professional, 259
Pareto, Vilfredo, 63
Pareto chart, 62, 63, 66–67, 83–84, 97, 98, 139
payback, 125
persistence, 24–25, 32
personal
characteristics, 177–78
development, 154, 177–78, 181, 182, 183
phantasmagoria, 70
pi, 47–48, 76
plagiarism, 230
plastics, 166
plow, the, 58, 166
positive attitude, 182
Prados, John, 64
prime rate, 125
printing press, 157, 164, 167
problem solving, 90–92, 93–94, 95–96, 144, 184, 187, 188, 189, 190, 196, 243
problems
environmental, 251, 253, 254
ethical, 243
social, 251, 253, 254, 255
procedures, testing, 129–30
product life, 254

production schedule, 107, 108, 109
professional
- behavior, 238–39, 244
- characteristics of a, 19
- organizations, 256, 259

professionalism, 40, 238, 239, 242, 244
professions, 19
professors, as a resource, 168–69, 176
profit and loss statements, 249
project planning, 109
proprietary information, 130
puzzles, creativity, 3

reasoning, 184, 187
Reid, Thomas, 241
Renaissance, the, 164
report writing, 130, 144, 145, 146, 147, 149, 150
resources, using, 168–69, 175, 176
respect, 127
responsibility, 146, 153, 183
return on investment (ROI) 125, 217
risk and return, 123, 124–25
roles, at work, 133
Roman developments, 162

safety, 261, 262, 263
- on the job, 42
- inspection sheet, 263
- rules, 262

scatter plot, 87, 88–89, 90, 136
science, 191
- function of, 195

scientific method, 60–61, 63, 64, 65, 66, 89
social
- ills, 255
- issues, 251, 253, 254

society, development of, 58
Society of Manufacturing Engineers, 256
solenoid, description of a, 52–53
spreadsheet, using a, 192
statistics, 205–06, 213
structures, organizational, 245, 248
student success, 21–23, 168–76
study groups, 130, 176
study habits, 122, 145
study skills, 80–81, 168, 171
success
- academic, 24
- in college, 168, 171
- on the job, 51
- personal, 24, 127

Swindoll, Charles, 182

taking notes, 168, 174
teachers, working with, 176
teams, 198–99, 200
- advantages of, 201
- development of, 198, 203
- in engineering, 198
- process of, 198
- roles in, 204
- types of, 198, 202, 207

teamwork, 198, 199
technician, function of, 195
technological age, 157
technology
- engineering, 14, 18, 195
- impact on civilizations, 57
- practice of, 191

telephone, 166
test results, 136–40, 143
- charting, 139
- compiling, 134
- documenting, 143

testing procedures, 129–30, 133
textbook
- marking a, 173
- using a, 168, 171, 173

time
- management, 168, 172
- robbers, 172
- to market, 107

trigger effect, 157, 166
triggers, 57, 58, 157, 166
trigonometry, 205, 211
Twain, Mark, 60–61

variable costs, 245–46, 249
variations, 130, 131
vocations, 19

Whitehead, Alfred North, 15–16
work journal, 100
writing
- in college, 230
- a cover letter, 146
- elements of, 231
- process of, 228
- memos, 146

writing skills, 146, 148, 149, 150